UNIVERSITY SPIN-OFF COMPANIES

University Spin-off Companies

*Economic Development, Faculty
Entrepreneurs, and Technology Transfer*

edited by
Alistair M. Brett,
David V. Gibson,
and
Raymond W. Smilor

Rowman & Littlefield Publishers, Inc.

ROWMAN & LITTLEFIELD PUBLISHERS, INC.

Published in the United States of America
by Rowman & Littlefield Publishers, Inc.
8705 Bollman Place, Savage, Maryland 20763

British Cataloging in Publication Information Available

Library of Congress Cataloging-in-Publication Data

University spin-off companies : economic development,
faculty entrepreneurs, and technology transfer / edited by
Alistair M. Brett, David V. Gibson, and Raymond W. Smilor.
    p.   cm.
Based on selected papers presented at a conference held
April 1988 at Virginia Tech.
Includes bibliographical references and index.
    1. High technology industries—United States—
Congresses. 2. New business enterprises—United
States—Congresses. 3. Industry and education—United
States—Congresses. 4. Research, Industrial—United
States—Congresses. I. Brett, Alistair M. II. Gibson,
David V. III. Smilor, Raymond W.
HC110.H53U55   1990
338'.06'0973—dc20   90-45866 CIP

ISBN 0-8476-7646-3 (alk. paper)

5   4   3   2   1

Printed in the United States of America

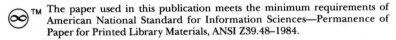

™ The paper used in this publication meets the minimum requirements of
American National Standard for Information Sciences—Permanence of
Paper for Printed Library Materials, ANSI Z39.48–1984.

Contents

v

Figures

Tables

Preface

Currently, the United States spends about $15 billion each year for basic research in our universities and government laboratories—about ten times more than any other nation spends or even has the in-place capability to spend. It has probably cost about a trillion dollars over the past 40 years to build this unique capability, which is the "seed corn" for the thousands of new products, processes, and services that ultimately result. This investment potentially provides the United States with a commanding advantage over all other nations. Unfortunately, this advantage has not been effectively realized.

Failure to capture the benefits of such an astonishing investment stems from a lack of incentives for further commercial development. Once a new discovery has been made in a basic research laboratory, about 90 percent of the ultimate cost, risk, and time to commercialize it still remains! Without incentives for such high-risk investments, a new discovery often can remain underdeveloped for long periods of time.

Moreover, the populist concept has persisted in the United States that federally funded technology belongs to all the people, and therefore cannot be exclusively owned by anyone. As a result, only about 4 percent of the 28,000 government patents have ever been licensed, since few companies are willing to pioneer new developments—at considerable cost—and then see their competitors follow at no risk. What is more, the 28,000 patents could have been 300,000, but there was no incentive to patent new discoveries. By freely publishing our basic research—without first patenting it—the United States has fueled much of the development work around the world for the past 50 years. Once patented, however, results can still be published as usual, and the future commercial rights will be protected, as well.

Fortunately, the Technology Transfer Acts of 1980, 1984, and 1986 have now authorized universities, government laboratories, and individual contractors both to own exclusively any technology developed under government funding, and to benefit directly from licensing or developing it further. All benefits flow back to them instead of into the federal treasury. Therefore, for the first time, incentives exist to patent and license technology.

Many universities have begun to take advantage of these new laws, generating substantial income for the university, the departments, and individual faculty members. Also, as industrial liaisons are developed, increased private-sector funding of basic research in universities almost always results as well. Clearly, the basic mission of the university needs to be protected under these arrangements. In this connection, it is important not to "rediscover the wheel," but to take advantage of systems already worked through by other institutions.

A national interest is involved here, as well. Industrial competitiveness in the global marketplace will be less and less dependent on the classical factors of comparative advantage (access to cheap labor, natural resources, arable land, capital, and propinquity to markets). Instead, comparative advantage increasingly will depend on knowledge-intensive value added factors (innovation, automation, and lifelong reskilling of the adult work force).

The great advantage that the United States has over other nations lies in this unique knowledge-intensive capability to generate next generation leading-edge technology and the flexible computer-integrated manufacturing facilities to make an ever-changing mix of new products. Effective technology transfer from universities to industry, therefore, is a vital requirement for global competitiveness, but it can be a win-win arrangement that also enhances the academic mission at the same time.

D. Bruce Merrifield
Assistant Secretary for Productivity, Technology, and Innovation
U.S. Department of Commerce
1988

Acknowledgments

This book is the result of the contributions of many people and organizations. We thank all the authors for their research and writing on the changing dynamics of regional and worldwide economic development through faculty entrepreneurship, technology transfer, and university spin-off companies.

We are most grateful to the RGK Foundation and its president, Ronya Kozmetsky, for helping to sponsor the conference on which this book is based. The staff of the Donaldson Brown Center for Continuing Education at Virginia Tech also deserves our thanks for providing the services to run the 1988 conference on the university spin-off corporation. Special appreciation goes to Lou Ann Phipps at Virginia Tech for her work in helping to run the conference and in collecting the presentations for inclusion in this book.

We are indebted to George Kozmetsky, director of the IC² Institute, which provided valuable support in the preparation of *University Spin-off Companies*. Rose R. Orendain of the IC² Institute deserves special recognition. She typed the manuscript through many revisions with terrific efficiency and good spirit. We also thank Linda Teague of the IC² Institute for preparation of the book's figures and tables.

We greatly appreciate the support and encouragement of Dean Robert E. Witt and Associate Dean Robert S. Sullivan of the Graduate School of Business at the University of Texas at Austin, as well as Professor James

S. Dyer, chairman of the Department of Management Science and Information Systems, and Professor Timothy W. Ruefli, chairman of the Department of Management. Finally, each of us wishes to thank his coeditors for an enjoyable and productive collaborative effort.

Introduction

Alistair M. Brett, David V. Gibson, and Raymond W. Smilor

In April 1988 Virginia Tech held a national conference on the university spin-off corporation. This book is based on selected papers presented at the conference. A spin-off company is defined as one that produces a product or service originating from research at a university. In many cases, the faculty member involved in the research will have started the company and may leave the university to run it, or an outside management team may be formed.

The Virginia Tech conference was planned to provide participants with an understanding of methods for commercializing university-related technologies through spin-off corporations. Presenters and participants were drawn from faculty and staff of U.S., Canadian, and European colleges and universities, industry, the legal profession, venture capital companies, state and federal agencies, and the business and industrial community.

During the past five years, several researchers in the United States, Canada, Sweden, and the United Kingdom have studied the university spin-off company phenomenon. In spite of some notable successes, such as Hewlett-Packard from Stanford University and Digital Equipment Corporation—one of more than 200 companies emerging from MIT—we have yet to find an efficient solution to the problem of launching academic ventures.

Many universities across the country are involved in creating new companies based on research results. Some of the reasons given for the recently increased spin-off activity include: promotion of a state's economic diversification, creation of local jobs, attraction and retention of quality faculty, social responsibility to turn ideas into useful products, and the generation of income.

Such spin-off firms are only a tiny fraction of the approximately 600,000 new companies being formed annually in the United States. However, with many states now expecting universities to be a greater force in economic development, university spin-off companies could be important catalysts for regional economic development. Further enhancing this viewpoint is the growing "thoughtware" economy and the increased involvement of universities in applying innovative solutions to problems in the manufacturing and service sectors.

Activities designed to stimulate technology transfer between the business and university sectors are rapidly increasing, both in the United States and other countries. In spite of this recent attention, there is little collected knowledge on the practical issues involved for universities and their business communities in the promotion of spin-off companies. A recent NASA Southern Technology Application Center report notes that there is a very wide range of legal-contractual agreements, foci, viewpoints, and degrees of success in arrangements to commercialize university-developed technologies.

The NASA survey—together with others—shows that there is considerable confusion between discussions of how much involvement universities should have in for-profit enterprises to commercialize technology, and technical considerations of how to set about the task. Some institutions are looking for instant revenue benefits, whereas for others the activity is a natural outcome of industry-sponsored research or industrial extension programs.

In discussing the intertwining of corporate and national competitiveness, the issue is the need not only to innovate, but to exploit innovation in the marketplace. It is only recently that many universities and individual faculty members have become convinced of the benefits to be derived from aggressive programs to commercialize the results of university research and development. Faculty attitudes are changing at universities of a variety of sizes and cultural orientations. Increasing interest in commercializing their R&D has led universities to develop new mechanisms that optimize the market potential of their research. The development of university spin-off companies is one such route. Furthermore, "the generation of knowledge is not usually considered a product in the same way that private industry regards the inventions of its scientists and engineers. . . . But the value of [this] asset, whether real or potential, is becoming so great and the need to use it so acute that approaches of the past . . . are no longer adequate to the task at hand" (Goldstein 1987).

Income to universities from licensing is low. For example, in a recent survey of NSF-funded projects, the probability of a significant patent income was found to be only 1.5 parts in 10,000 person-years of research effort. While much more study is needed, it appears that the probability of significant wealth creation is higher when a new spin-off company is created around the technology. The latter is true especially when the invention in question is not a (rare) major breakthrough.

The importance of technological change on economic growth, of R&D spending on increased productivity, and of quality education in promoting growth and productivity has been well studied in the past 20 years or so. More recently, the 1985 Presidential Commission Report on Industrial Competitiveness stressed the need for science and technology to be relevant to commercial uses. Linkages between universities and industry have a long history. It has been suggested, however, that the rapid technological change experienced in the past decade "generates interpenetration of the academic and industrial domains due to an increase in the market value of many university products and a blurring of the lines between basic and applied research." Consequently, "both sectors become interested in the same research talent and both are motivated to profit by the commercialization of university resources" (Baba 1987, 191).

The University's Role

Discussions over increased faculty and university involvement in the commercializing of research and the marketing of related services such as problem-solving and consulting will no doubt continue. There are some who express concern that the traditional open communication of research results will be diminished or that an emphasis on applied research could hurt scholars engaged in basic research. Others point out that technology transfer programs can help retain top-quality faculty and that the overall level of research commercialization is not significant enough to tempt universities to change their policies radically.

These trends are not restricted to the United States. In the United

Kingdom, new government policies call for greater emphasis on technology transfer between educational institutions and industry. In Sweden, some 90 companies—about 30 of which manufacture products—have been spun-off from Chalmers University. Other universities in Canada and most of the European Community countries are actively supporting the creation of spin-off ventures. In Europe too, the concern has been expressed over the need to "minimize the consequences of possible conflict between academic and scholarly imperatives and those of business and commerce, as links with industrial research interests and sources of funding become increasingly important to institutional viability" (Taylor 1986, 13).

A 1985 report from the (London) Economist Intelligence Unit in discussing the European situation, noted: "Higher education has played, and will continue to play, a significant role in the creation of economic wealth via the formation of new ventures." But the report goes on to note that "exploiting technology through the formation of a new company is probably the most complex form of university based technology transfer" (*Universities and Industry* 1985, 61).

Overall U.S. corporate spending for R&D is moving upward, but the rate of increase has slowed. The United States significantly trails Germany and Japan on nondefense R&D as a percentage of GNP. During calendar year 1988, total U.S. expenditures on R&D were around $132 billion—an increase of some 7 percent over 1987 estimates. Estimates vary, but U.S. colleges and universities account for some 12 percent, or nearly $16 billion, of these expenditures.

According to a recent Battelle report, "as greater emphasis is placed on issues of competitiveness and the most efficient use of technical resources, a greater effort should be directed toward understanding the role that university research plays in the national economy and the incentives and barriers which influence utilization of this resource" (Battelle Institute 1987, 15).

The common theme connecting the chapters in this volume is that the university research base must be strategically managed to act as a more effective source for new business start-ups. Furthermore, while the research enterprise becomes more sensitive to the value of its products, the driving function of a university to teach the known and explore the unknown cannot be distorted. Industrially sponsored research now accounts for some 6 percent nationally of research funding at universities, and has not adversely affected basic research or teaching. Indeed, it has been the contrary. So, too, we expect research commercialization through new venture formation to move universities into a new and important societal role.

The chapters that follow represent different approaches to commercializing technology through spin-off ventures at a number of universities. They also raise several problems and challenges. We believe that the nation's universities are capable of overcoming the constraints and capitalizing on the challenges to increase industrial competitiveness, create wealth as well as improve the quality of life of the community, provide greater returns on our investment in higher education, and enhance the traditional knowledge-generating role of the university sector.

References

M. L. Baba, "University Innovation to Promote Economic Growth and University/Industry Relations," in *Technological Innovation and Economic Growth,* P. A. Abetti et al. eds. (Austin: IC² Institute, University of Texas, 1987).

Battelle Institute, *Probable Levels of R&D Expenditures in 1988: Forecast and Analysis* (Columbus, Ohio: Battelle Institute, 1987).

M. B. Goldstein, "Equity Financing: Research Partnerships," in *Financing Higher Education: Strategies after Tax Reform,* R. E. Anderson and J. W. Meyerson, eds. (San Francisco: Jossey-Bass, 1987).

W. Taylor, "Crisis in the Universities," *OECD Observer* 143 (November 1986).

Universities and Industry, Economist Intelligence Unit Special Report No. 213, I. MacKenzie and R. R. Jones, eds. (London: Economist Publications, 1985).

Part I

The University and Economic Development

1

Global Economic Competitiveness and the Land-grant University

John E. Cantlon and Herman E. Koenig

Land-grant universities were established in Michigan and Pennsylvania in 1855 because the universities of the day chose not to accept the challenge of bringing the emerging discoveries from the sciences to bear on social and economic development. These two institutions were not inventions of academics; they were the outgrowth of public frustration with the growing isolation of academe from the greater social and economic concerns of the time. Today, many detect a similar public frustration with the land-grant institutions themselves, which have failed to keep faith with their explicit mission.

Students of Michigan State University's history know that, 15 years after the institution opened, there were farmers as near as 15 miles away who were unaware that: (1) there was an agriculture college near Lansing, and (2) they might benefit from the presence of such an institution. The small stream of graduates and published articles leaving the fledgling institution made little early impact on the then largely agrarian society's needs. It was not until the agriculture college faculty began to work with farm groups and county leadership to identify common priorities that the great idea of the land-grant institution began to make a difference. Subsequently, political support both in Michigan and nationally resulted in the Hatch Act (1887) and later the Smith–Lever Act (1914) which provided funding for state land-grant colleges with explicit applied research and cooperative extension missions in agriculture and the mechanical arts.

The power of these well-coordinated private sector–government–university economic development machines led to the very productive U.S. agricultural technologies that now swamp the United States with surpluses, costly farm subsidies, and farmers struggling to compete adequately in today's international market.

But small and medium manufacturers, serving as suppliers to the nation's manufacturers of mechanical equipment, have had no formal structures to link the emerging discoveries of science and engineering to their needs. Even in the realm of agriculture, the land-grant system was far more proficient in dealing with technical issues than with the many other aspects of competitiveness in the rapidly changing international, and regional political, economic, and social environments—such as marketing strategies, logistics, product differentiation, quality and cost control, finance, and externalities such as environmental impacts and the displacement of farm workers.

All of this is history—history that began well over 100 years ago. Not only has the world changed dramatically in that time, but also the rate of change is accelerating. What are the economic development issues of Michigan, the North Central states, the United States, and the world; and how can or should land-grant universities respond more effectively?

The basic thesis of this chapter is that the fundamental economic issues we face today and in the years ahead have to do with innovation, quality and cost control, entrepreneurship, and all that these imply socially, culturally, and technologically. This includes particularly the ability to recognize that change—not stability—is and will be the norm of the future, and that the ability to embrace change as opportunity—rather than threat—is the key to economic security. The objectives of this chapter are to (1) develop this basic thesis and (2) show how Michigan State University (MSU) as an AAU/land-grant university is attempting to modernize its land-grant mission.

The Dynamic State of the Global System

By and large, humankind on planet Earth now lives in a global economy, a global ecosystem, and an increasingly homogenized global culture. From a fundamental point of view, economic statistics such as exchange rates, stock markets, trade balances, or even GNP figures are only indicators of activity within this planetary system. To understand how this system really functions we need to look at the stuff of which the system is made—its ingredients. As fundamental ingredients, one can list:

1. The human population (consumers and human resources);[1]

2. Knowledge and technology;

3. Biological and ecological resources;

4. The physical environment, including material and energy resources;

5. Production capacity (agricultural, aquacultural, and physical plant assets and essential services);

6. Physical transportation capacity;

7. Physical communication capacity (information-generating, storing, networking, and accessing capacity);

8. Cultural mechanisms for the exchange of capital and consumer goods and services.[2]

The first remarkable thing about a global system is that—except for the daily flux of solar energy that impinges on the earth, and that drives the natural and domesticated biological systems—this system is closed with respect to the eight ingredients above, which we therefore call the "states" of the system. The second remarkable thing about the global system is that, in principle, its future states at any point in time are inextricably dependent on the states today and at previous points in time.

The third remarkable thing is humanity's unique ability as a species to learn of or make assumptions about parts of this enormously complex system and to modify the states and their interrelationships for both individual and collective purposes. The capacity of the human participants to intervene in this manner, often in large numbers and over short periods of time, makes prediction of the dynamics of the system at best a risky art.

It is a most fundamental principle that there is no such thing as a stable global economy. Change is intrinsic to the system. The process of human learning begets new knowledge; and knowledgeable people both generate more knowledge and train more knowledgeable people, most of whom use parts of this ever-expanding knowledge in countless ways to modify the states of the system and their interrelationships.

The states of the global system are interdependent and tend to change exponentially. Figure 1.1 illustrates, on an evolutionary time scale, the exponential growth in the world's human population and its correlation with fossil energy consumption per year—one of the critical material states (ingredients) of the contemporary economy. Figure 1.2 illustrates the growth in the annual consumption of this ingredient on a time frame

Figure 1.1   Energy Consumption (BTUs) and Population Trends from 0 A.D. to the Present

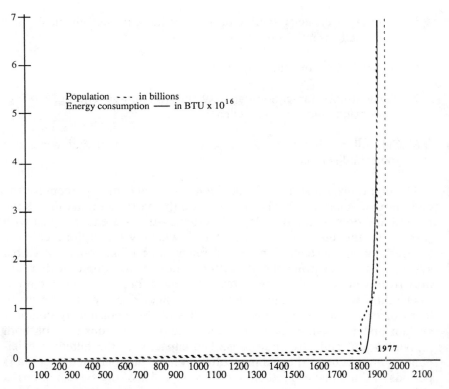

Population - - - in billions
Energy consumption ——— in BTU x 10^{16}

Source: E. Cook, "The Flow of Energy in an Industrial Society," *Scientific American,* 224, 3, (September 1971), pp. 134–44.

Figure 1.2   Cycle of World Oil Production Plotted on the Basis of Two Estimates of the Amount of Oil That Will Ultimately Be Produced

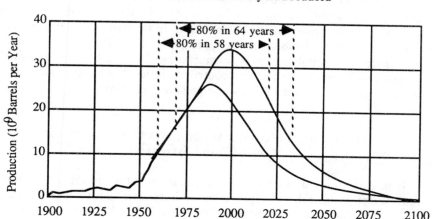

Source: Hubbert M. King, "The Energy Resources of the Earth," *Scientific American,* 224, 3, (September 1971), p. 69.

appropriate for strategic planning, along with an upper and lower bound on the projected state of petroleum production in the decades ahead. If we had the capacity to measure the rate of the accumulation of bits of information in our global cultures, its rate of increase might be even more dramatic.

Individuals, organizations, and nations struggle to maintain and improve their standard of living in the midst of exploding world population, new knowledge, changing energy and material resource bases, and local and global impacts of their collective action. Their ability to do so depends critically on the generation of new scientific knowledge and the manner in which it is developed and deployed through countless social, political, economic, institutional, and other cultural adaptations. In this respect it can be said that, in the final analysis, generation and effective deployment of knowledge and information constitute the engine of economic growth and development. Since knowledge grows and propagates geometrically, one or a few of the planet's cultures can no longer find security in preserving the past unless they are powerful and committed enough to also suppress the generation and deployment of knowledge by their global competitors. Security can only be found in the ability of individuals, organizations, institutions, and cultures to embrace change and capitalize on the opportunities it provides. And this is what innovation and entrepreneurship are all about.

It is illuminating to note the global economic power of modern Japan in contrast to the global military power of the United States and the Soviet Union. In a global economy with a relatively free flow of goods and services, world power based on industrial innovation and entrepreneurship appear to hold more long-term promise than military power based on costly and sophisticated war machines that must be continuously updated technologically and that contribute little to national wealth unless in fact they are used for military conquest and subjugation.

Innovation, Entrepreneurship, and Competitiveness in a Global System

It is clear that the rate of increase and use of scientific and cultural knowledge in our time is unprecedented in human history. It is also true that more than the soaring power of the minds of a select few in our research laboratories was needed to produce the thrill and luxury of contemporary life. Technological advances also require entrepreneurial and managerial skills and creative and motivated human hands. They invariably involve social and cultural adaptations in virtually all aspects of life, including specifically the culture of the integrated work places and management of our corporate enterprises. And this brings us to the first central issue on which to focus.

The United States has developed and sustains the most extensive higher educational system and the most powerful basic-research machine in the world. The Great Lakes region—in particular—has more large, high-quality universities and other postsecondary institutions than most other regions of this nation or of the world. Large universities in the Great Lakes states—most of them primarily publicly supported—rank among the nation's best. They engage in a broad spectrum of scientific research and creative endeavors, and they are as aggressive as any in transferring new knowledge to the commercial sector and in providing expert services and assistance to the public and private sectors. They have served as a nationally and internationally recognized source of well-educated and highly trained young men and women in the basic sciences, arts and humanities, engineering, management, law, medicine, and other professional fields. Yet in the realm of global commerce, the United States and the Great Lakes states have become less and less competitive since the 1960s, even with respect to the technological newcomers among nations. Why is this, can it be turned around, and what will it take to do so?

In an open competitive economy, the window of opportunity through which any individual or corporation can capitalize commercially on new technological advances is limited. It is an historical fact that, following World War II, the Midwest generally failed to move in an aggressive and sustained enough way to capture the commercial benefits that ultimately resulted from the flood of new technologies—multispectral sensing, solid-state electronic devices, telecommunications, computers, nuclear fusion, new materials, and so forth—flowing from the nation's universities and research laboratories (even the region's own). And the major firms failed to recognize the potential of incorporating new management philosophies—such as participatory management—in the organization and operation of their businesses, even though a few small firms in the region were demonstrating their effectiveness. Further, corporate leadership placed near-term financial gain ahead of winning the long-term competitive edge, which reduced research and stifled innovation. New business ventures based on new technologies were developed by entrepreneurs in other regions of the country and the world. In turn, the core of the new industrial activities in other regions of the nation served as career magnets for the top graduates from Midwest universities, thereby compounding the problems of the region with decades of "brain drain."

Application of some of these technologies to manufacturing processes in Michigan and the Midwest are only now being developed. But the lag in competitiveness of U.S. manufacturing businesses is clearly not confined to the production process, as many have believed. Innovation in production—as much as it is needed—will not fundamentally change the global competitive position of traditional U.S. manufacturing, as many had hoped.

It is an historical fact that even in the realm of established Midwest product lines such as automobiles, for example, almost all recent innovations—the radial tire, the disk brake, computer-controlled electronic ignition, the stratified charge engine, four-wheel steering, and others—were first mass-produced by offshore competitors. It is also an historical fact that these same competitors adopted rigorous quality and cost controls and numerous new marketing, organizational, and management innovations that provided new levels of market differentiation. Yet both the scientific and engineering principles and the management concepts on which these innovations are based have been known for many years—many of them being products of U.S. researchers and scholars.

Another fact is that the U.S. government and even the state and local governments had developed a largely regulatory posture vis-à-vis industry. In contrast, our most aggressive international competitors enjoyed a much more collaborative and accommodating relationship with their governing bodies. Perhaps the ultimate irony is that heavy U.S. assistance in the postwar recovery in Europe and Japan encouraged a government–industry collaboration that our own laws prohibited. A culture that places the individual ahead of the whole tends to be predominantly regulation-minded relative to business and industry, and costly litigiousness and risk avoidance burden the competitive enterprise with layers of inefficient, defensive management structures. However, this in itself does not explain why Midwest industries have not been as aggressive and effective as even their domestic competitors in capturing the benefits of new knowledge and technology flowing from the laboratories of the world.

The fundamental issue is that we have now entered a new economic era—an era in which, for firms of the mature industries,[3] catching up with the competitors' technologies and/or management and marketing strategies does not constitute success. The competitor can only be overtaken by leapfrogging to new opportunities opened up by new market knowledge, product differentiation, new and improved product and production technologies, and new management/organizational innovations in the work place and the boardrooms. These are now the essentials for successful global economic competitiveness.

Global Economic Competitiveness and the Research University

The economy of the Midwest, in particular, is currently in a transition from one based on relatively well-established and secure production for largely domestic markets of traditional manufactured products and agricultural commodities, to one based increasingly on the production and marketing of product lines that have a much higher technology content and rigorous quality standards and are finely tuned and aggressively marketed to discriminating and changing national and international markets. This transition has been underway since the 1980–81 recession, but will be long and difficult. It will require cooperative efforts between the private sector and all levels of education from kindergarten through twelfth grade to the large research universities, not-for-profit R&D institutes, and all levels and branches of government. This requires substantial cultural reawakening, and in the process we must reestablish a trust in each other so that together we can compete more successfully in the intensifying global competition. To continue our past behavior will most assuredly lead to further economic, technological, and cultural weakness with resultant risk for the nation. The transition that must be expedited in the decades ahead is circumscribed by a number of issues over which the individual, the firm, and the universities have little direct control. They include, but are not limited to:

• Growing need to address global CO_2 acid deposition and other circulating pollutants;
• Growing sophistication of developing nations to generate, acquire, and use sophisticated information and technologies;
• Growing recognition by both superpowers that projection of economic power may produce greater societal benefit, incur lower risks, and have higher survival value than projection of military power;
• Growing stress on remaining benign global energy resources;
• Growing threat of extinction of global genetic and ecological information before we can decode and examine its potential uses; and
• Growing threats of social instability from ever more aware but economically and culturally deprived populations.

We believe, however, that one major issue—and one that we in higher education can and must do something about —has to do with improving the linkages connecting academe, the private sector, and government. Academe is the source of much of the new basic knowledge and understanding. The public and private sectors of our economy are where new knowledge is translated into competitive new and improved products and services, and where the products of the educational process are transformed into practitioners in a sociotechnological culture capable of competing economically, technologically, and culturally with other cultures of the globe. In 1990, 74 percent of the U.S. population are in the 25–54 age bracket; and today's work force will constitute 85 percent of the work force by the turn of the century. Given this, one can assume that an even greater percentage of tomorrow's managers are already at work.

The following sections will use a series of diagrams and illustrations to show how MSU as an AAU/land-grant university has been attempting to redefine its role in and strengthen its linkages with some of the priority sectors of the Michigan economy. Similar operations are underway in other universities throughout the nation, encouraged in part by federal, state, foundation, and private-sector incentives.

Figure 1.3 conveys the nature of the support system for a typical AAU/land-grant research university like MSU. This system is a post–World War II product. The federal government provides the lion's share of basic and applied research funding; the state government and student tuition and fees (roughly ⅔:⅓) pay for instruction; and a mixture of federal,

Figure 1.3   A Model of the U.S. Research University's Support Systems

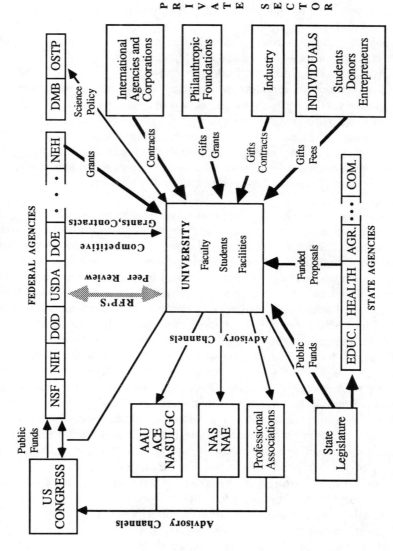

Source: The authors.

state, and local governments and the receiving clients paying for access to expert services.

At the core of the U.S. graduate research/land-grant university is the faculty of experts in the various disciplines, professions, and fields of study that undergird our complex culture. This faculty of experts provides the research, instruction, extension, consulting, and other outreach/inreach activities according to institutional and disciplinary traditions. The system's organization, however, was not designed to coordinate faculty expertise in support of specific economic development goals. Except possibly for some faculty in the colleges of agriculture and medicine, and occasionally in the schools of business and engineering, their clearly articulated primary mission is to advance knowledge and understanding in the disciplines and professions and to instruct graduate and undergraduate students in these disciplines.

The basic structure of concern here is portrayed in simplified form in the three columns of Figure 1.4. The column on the left lists some of the major sources of funding that enable the faculty to search for new knowledge and understanding. The middle column lists the organization of the various academic disciplines and schools around which faculty do research, creative scholarship, and teaching activities and around which programs and faculty are developed, organized, funded, and rewarded. The column on the right lists the basic economic sectors around which the world's economy, governance, and culture are organized.

The federal sources of university R&D funding such as the National Science Foundation and the National Institutes of Health are highly selective in the academic disciplines they support. Except for general legislative appropriations to higher education, this is also true for state R&D funding, such as Michigan's Research Excellence Fund.

The entities in the academic resources column are not one-to-one counterparts of those in the economic sectors column. It is often said that, while business and industry have problems and opportunities, universities have academic departments and intellectual disciplines. Problems are rarely single-discipline matters, and addressing them requires that many departments work in concert. What is required, then, are organizations and procedures that can provide more effective interaction between faculty and facilities within the university and practitioners in the economy. MSU has tried to do this in four ways:

1. Provide centers, institutes, and other "bridging structures" within the university, where discipline-oriented faculty and facilities can focus their collective talents on economic opportunities that are

Figure 1.4   Organization of Business and Academe, with Example Sources of Academic Funding

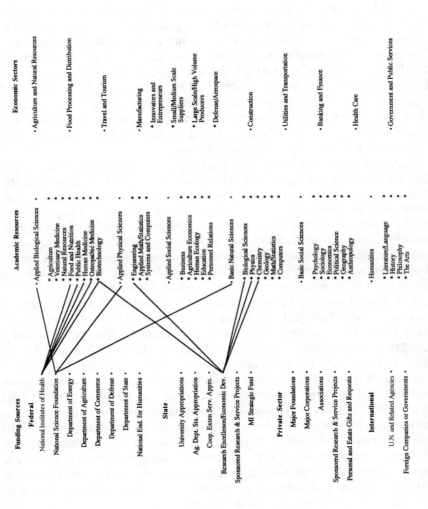

Source: The authors.

more compatible with and accessible to practitioners in business and industry;

2. Establish statewide networks with sister institutions of higher education and private- and public-sector partners to promote and support innovation and entrepreneurship in existing business enterprises;

3. Promote and facilitate new spin-off companies and establish special arrangements directly with private-sector partners to promote and facilitate commercialization of promising technologies and concepts; and

4. Make adaptations in institutional policies and educational programs to help keep them tuned to changing technological and economic competitiveness issues.

Bridging Structures within the University

The bridging structures currently in existence at MSU are listed in the middle column in Figure 1.5. The most important and best supported are the long-established Agricultural Experiment Station and Cooperative Extension System that link MSU's basic and applied R&D in agriculture and natural resources, basic natural sciences, engineering, applied biological and social sciences, and other fields to the agricultural and natural resource sectors of the economy, as illustrated by the connecting lines.

More recently the Food Industry Institute was established to provide a more sharply focused bridge between the food processing and distribution sector of the economy and the major academic fields of knowledge—identified by the connecting lines—on which successful firms in this sector of the economy are increasingly dependent.

One of the specific functions of the Food Industry Institute is to establish a working relationship between industrial practitioners and an appropriate team of academic specialists as required to identify problems and opportunities, to conduct problem-oriented R&D, and to help otherwise anticipate opportunities and be responsive to the specific needs of the food industry. This orientation is typical of many of the bridging structures within the university.

Figure 1.6 illustrates the Division of Engineering Research as a long-standing R&D link between the engineering department and—more gen-

Figure 1.5   University-based Examples of Bridges between Business and Academe: The Agricultural Experiment Station, and the Food Industry Institute

Academic Resources

- Applied Biological Sciences
  - * Agriculture
  - * Veterinary Medicine
  - * Natural Resources
  - * Food and Nutrition
  - * Public Health
  - * Human Medicine
  - * Osteopathic Medicine
  - * Biotechnology

- Applied Physical Sciences
  - * Engineering
  - * Applied Math/Statistics
  - * Systems and Computers

- Applied Social Sciences
  - * Business
  - * Agriculture Economics
  - * Human Ecology
  - * Education
  - * Personnel Relations

- Basic Natural Sciences
  - * Biological Sciences
  - * Physics
  - * Chemistry
  - * Geology
  - * Math/Statistics
  - * Computers

- Basic Social Sciences
  - * Psychology
  - * Sociology
  - * Economics
  - * Political Science
  - * Geography
  - * Anthropology

- Humanities
  - * Literature/Language
  - * History
  - * Philosophy
  - * The Arts

Bridging Structures Within the University

- Office of Research Development
- Lifelong Education Programs
- Cooperative Extension Service
- Agricultural Experiment Station
- Division of Engineering Research
- Kellogg Biological Station
- Technology Transfer Center
- Food Industry Institute
- Industrial Development Institute
- Center for Environmental Toxicology
- Michigan Travel, Tourism and Recreation Resource Center
- Center for Composite Material and Strategy
- Electronic R&D Laboratory
- Food and Pharmaceutical Packaging Program
- International Business Development Program
- Advanced Management Program
- Materials and Logistics Management Program
- Center for Redevelopment of Industrial States
- Social Science Research Bureau
- School of Labor and Industrial Relations Practitioners Program
- Labor Program Services Personnel Management Program Services
- Center for Communication Research
- Center for Remote Sensing
- Center for Cartographic Research and Space Analysis

Economic Sectors

- Agriculture and Natural Resources
- Food Processing and Distribution
- Travel and Tourism
- Manufacturing
- Innovators and Entrepreneurs
- Small/Medium Scale Suppliers
- Large Scale/High Volume Producers
- Defense/Aerospace
- Construction
- Utilities and Transportation
- Banking and Finance
- Health Care
- Government and Public Services

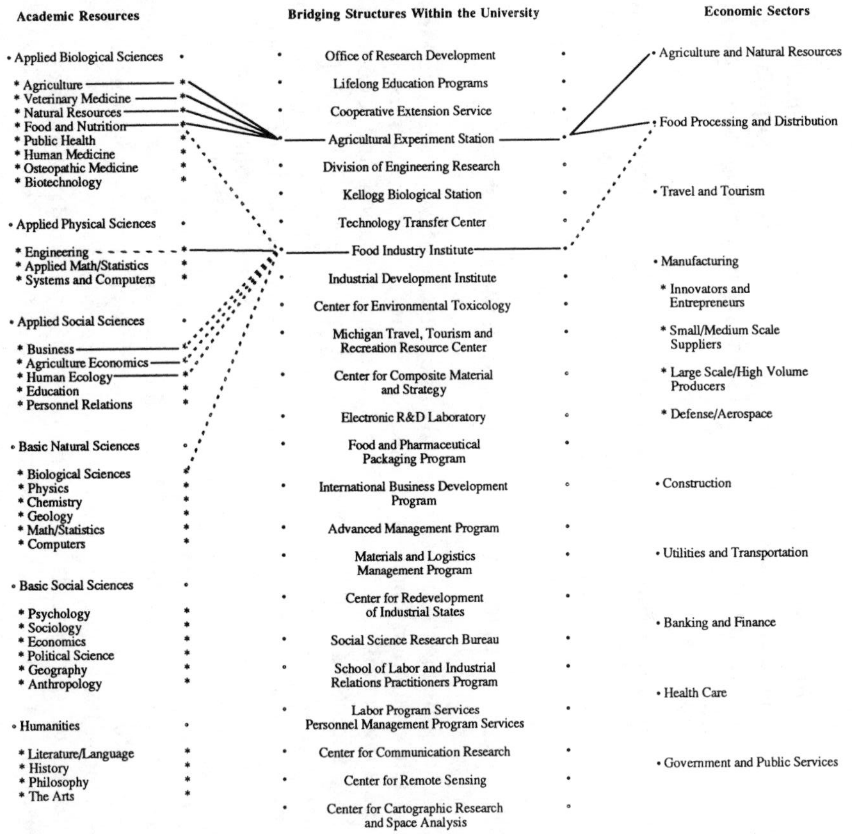

Source: The authors.

Figure 1.6   University-based Examples of Bridges between Business and Academe: The Division of Engineering Research, the Industrial Development Institute, and the Advanced Management Program

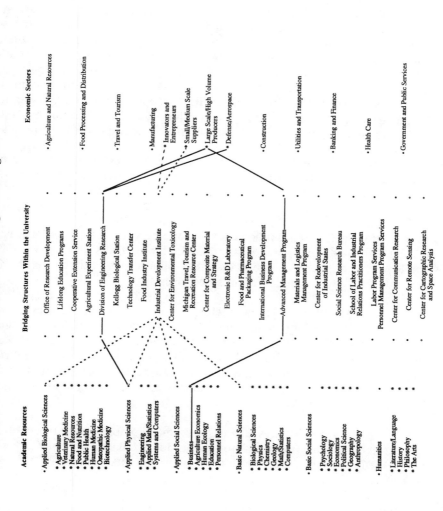

Source: The authors.

erally—the applied physical sciences of the university, and large indus-trial corporations and the defense and aerospace industries. Through completely different and independent channels, such as the Advanced Management Program and the Materials and Logistics Management Pro-gram, the College of Business provides these same firms with the newest concepts and skills in marketing, logistics, and other areas of business management. Many corporations are able to identify, effectively inte-grate, and utilize the two fields of knowledge (business and engineering) internally, using their own team of specialists. A substantial part of MSU's sponsored research and services funded by private sources comes from large corporations. Small and medium-size firms, on the other hand, frequently do not have this internal capability. Recognition of this has led to their designation as priority clients—thus best applying the university's limited resources.

MSU believes that young, technology-driven smaller companies are likely to be the key to Michigan's economic development, diversification, and competitiveness for four reasons:

First, statistics tell us that small and medium-size firms have accounted for the entire net increase in jobs in this country since 1981. The forecast is that this will continue. Small manufacturing firms (less than 100 employees each) provide more than 4.5 million jobs and represent 94 percent of the U.S. manufacturing enterprises.

Second, many of these young, technology-driven companies operate in the shadow of university campuses, where they can help rearticulate the university with the nation's economy. They also provide opportunities and challenges that are helpful in attracting and keeping high-quality faculty, provide practical work experiences for faculty and students, and provide training opportunities or adjunct faculty appointments for the firm's employees.

Third, in the United States at least, new small companies are frequently quicker at converting research into new products and services than larger, established, and more risk averse firms with high internal inertia. This is critical to U.S. global competitiveness because our rivals are not beating us in research and discovery; rather, they are beating us in the speed with which they move laboratory discoveries to products that continue to compete in the marketplace.

Fourth, a growing number of large, well-established corporations are looking more and more to forming strategic partnerships with promising young high-tech firms. The new companies can target early small markets and are willing to take risks. Moreover, they offer promising new products as well as a high level of creativity and maneuverability. The large corporations offer massive marketing and financial resources and access to large-scale production. Through strategic partnerships—rather than buyouts—each can continue to complement the other.

The more than 15,000 small and medium-size manufacturing firms in Michigan present both a special challenge and an opportunity for state government and the institutions of higher education in the state. They are typical of small firms throughout most of the Great Lakes states. Most are suppliers to original equipment manufacturers (OEMs) and have been isolated from both researchers and diverse consumer markets for many decades, yet they are now suddenly being called on by the OEMs to achieve new levels of quality and cost control, participate in new computer-based logistical systems, and assume new levels of responsibility in component innovation and engineering. Most are ill prepared philosophically, technically, and organizationally to respond.

Perhaps the best measure of the technological immaturity of these firms in the Midwest is contained in the statistics now available from the NSF's Small Business Innovation Research (SBIR) program. Nationally, in the first four years of this program (1982–86), more than 35,000 proposals were submitted by small companies to the 12 federal agencies that conduct the applied research competitions required by law. Small firms in ten states received 69 percent of the total awards (more than $650 million), while the bottom 25 states—which include Michigan and most other Midwest states—received less than six percent of the funds awarded. Michigan, for example, has the same index (weighted average of population, wealth, and labor force) as Massachusetts yet received less than 12 percent as many awards in dollars.

Michigan's technically retarded population of small and medium-size firms with strong university connections is the critical link between the universities and Michigan's industry; and we consider these firms our priority clients, including new start-ups. Creation of MSU's Industrial Development Institute, highlighted in Figure 1.6, and aggressive support of the Michigan Technology Council and the SBIR program are but a few of the specific steps that have been taken to strengthen our linkages with them.

The central feature of MSU's Industrial Development Institute is its ability to: (1) promote and facilitate innovation and entrepreneurship in small and medium-size firms; and (2) organize specialized knowledge and expertise into programs of customized assistance to support them. The MSU College of Business is establishing a Center for Entrepreneurism and an International Business Development program that will also soon serve as focal points for economic development.

A complete "wiring diagram" showing how each of the bridging elements within the university links its various academic fields to specific sectors of the economy would be as intimidating and confusing to the reader as the university itself can be to the practitioner looking for access to its resources. The Cooperative Extension Service (CES), highlighted in Figure 1.7, was developed nearly a century ago to alleviate that problem for farmers and others in the agriculture and natural resources sector of the economy. CES has long since extended assistance to firms in food processing, packaging, and distribution and in travel and tourism; it assists and participates in county and regional economic development and planning councils, as well. More recently with funding from the Michigan Department of Commerce, MSU established the Technology Transfer Center to serve as a single point for accessing university resources on a universitywide basis (Figure 1.7). In fact, through a network of such centers at four of our sister institutions, the practitioner has direct access to the collective resources of all the participating universities.

Statewide Transfer Networks

Most of the 50 states, through their executive and/or legislative branches of government, have developed some sort of strategic plan for economic development—presumably building on their areas of relative strength. A recent study commissioned by the Michigan Strategic Fund, for example, identifies four fields of technology—automated manufacturing, advanced materials, biotechnology, and electronics and information—as strategically important to Michigan. The study also identifies three stages or levels of technological evolution in each of these strategic areas. The results are summarized in Table 1.1. The Michigan legislature has established a special fund called the Research Excellence Fund (REF), which provides $27 million in annual R&D grants to its state universities in support of emerging technology developments in the first three of the four areas listed in Table 1.1.

Through such sources of funding as oil and gas lease revenue from public lands that the state maintains, the Michigan Strategic Fund partially supports three freestanding, not-for-profit institutes of excellence that are close to but independent from university centers: the Industrial

Figure 1.7   University-based Examples of Bridges between Business and Academe: The Cooperative Extension Service, and the Technology Transfer Center

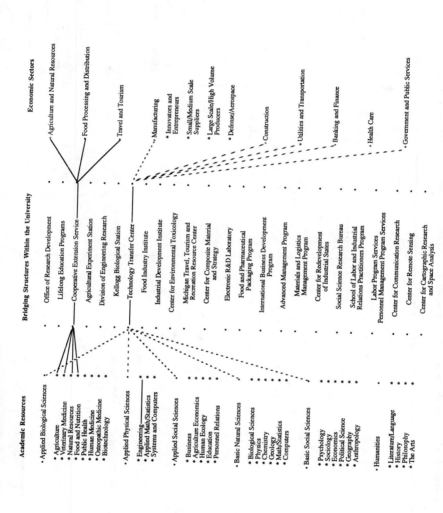

Source: The authors.

Table 1.1   Framework of Technical Assistance in Support of Economic Development in Michigan

Technological Areas of Strategic Value to Michigan	Transitional Technologies -- Proven technologies that are generally available and in use by more progressive firms in the industry	Cutting-Edge Technologies -- Technologies that are developed and being tried by only a few sophisticated firms in the industry	Emerging Technologies -- Technologies still under development and will be important in the longer term
Automated Manufacturing	• Computer-numeric-controlled machines • computer aided design • statistical process control	• CAD-CAM systems • flexible production cells • group technologies • Sensor-based robotic assembly and inspection stations	• artificial intelligence, robotic assembly stations • diagnostic tools driven by three dimensional machine vision
Advanced Materials	• alloys • engineered polymer	• reinforced-resin-based components	• ceramics for engines, armor power systems and tooling
Biotechnology	• plant tissue culture • improved seeds • micropropagation techniques for plants	• genetic engineering for microbial	• biotechnologies based on monoclonal antibodies and recombinant DNA
Electronic and Information	• semiconductor • microprocessors • computers • communication equipment	• custom designed microprocessors • integration of vision systems, effectors, robots, cell-level process controls, and real-time factory floor management	• parallel and network computer processing • high-density, high-speed microprocessors • electrooptical systems • artificial-intelligence-driven computer systems

Source: Developed from information contained in "Technology and the Michigan Economy: Elements of a Strategy," Summary Report, October 1987, Prepared for the Michigan Strategic Fund by the Center for Economic Competitiveness, SRI International.

Technology Institute (ITI); the Michigan Biotechnology Institute (MBI); and the Michigan Materials Processing Institute (MMPI).[4] They help to facilitate the movement of research discoveries to cutting-edge technologies and into commercial application in each of the first three respective areas listed in Table 1.1. These not-for-profit institutes complement the university-based institutes as sources of expertise in cutting-edge technologies and as sites appropriate for proprietary research.

The not-for-profit and university-based institutes and centers serve as resources to three complementary statewide transfer networks, as illustrated in Figure 1.8. Each of the three transfer networks focuses on selected sectors of the economy and/or levels of knowledge and technology so that together they serve the highly varied needs of the entire population of small and medium-size firms in the state—many of which are technologically ill prepared, as indicated earlier. Each network deals in an integrated manner with the technologies of production, human resources development, and adaptations in the culture of the work place as well as with the new management, logistics, and marketing perspectives often required to compete in the marketplace. And each works cooperatively through special arrangements with public- and private-sector professional associations, industrial consortia and associations, and community economic development alliances and corporations in carrying out their mission. The Michigan Modernization Service (MMS), for example—funded and managed through the Michigan Department of Commerce—was created to help small to medium-size manufacturing firms with immediate practical assistance in implementing state-of-the-art and transitional technologies. Since these technologies are already proven and used by advanced firms in the manufacturing industry, they (appropriately) draw their expertise primarily from private-sector firms and consultants, community colleges, and ITI—rather than from faculty of the research universities.

But if Michigan and other Great Lakes states are to move beyond a "catch-up" or survival position in the new global economy, they must also support their more technologically advanced firms in planned innovation so that they can leapfrog through the new windows of opportunity that only emerging knowledge and technologies can provide. Additionally they must help the capable among the less advanced grow to this position. The Technology Transfer Network (TTN), developed and operated as a joint venture between five major research universities of the state—Michigan State University, University of Michigan, Wayne State University, Michigan Technological University, and Western Michigan University—and the Michigan Department of Commerce, was designed and

Figure 1.8   Growth Stages of Knowledge and Technology and Primary Sources of Expertise

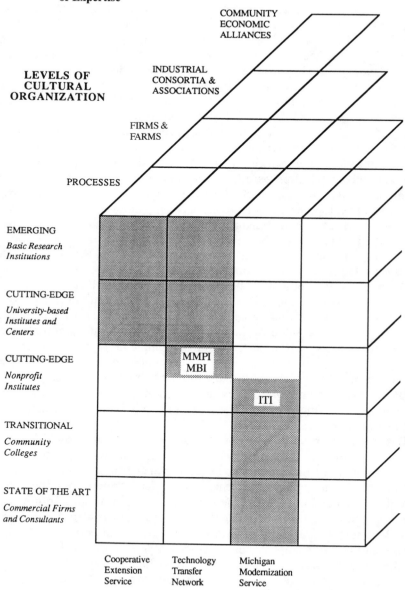

Source: The authors.

developed to provide such assistance. The network especially promotes and facilitates direct and ready access by firms to university sources of cutting-edge and merging technologies and advanced management/organization concepts. At MSU it works in concert with our Industrial Development Institute and the Food Industry Institute to establish stable working relationships with the firm. This relationship typically starts with a careful and responsible assessment of the firm's needs and aspirations in both the short and long terms and then builds incrementally over time toward productive and lasting working relationships with appropriate university faculty experts and/or university-based institutes and centers. These activities might include, for example, development of joint proposals for R&D funding under the federal SBIR program.

New start-ups and spin-off companies from the university—as important as they are—are not sufficient in Michigan and much of the Midwest. Only by expanding Michigan's base of industries that habitually maintain a working relationship with its universities (just as progressive farmers have done for a century) can the state overcome some of its most serious industrial limitations, diversify its economy, and in the long run measure up competitively with other states and countries of the world, such as Massachusetts, California, North Carolina, Europe, Japan, and the Pacific Rim. This is a task in which the TTN plays an important linkage role.

Historically, the Cooperative Extension Services (CES) of the land-grant universities of the nation have been effective instruments of technology transfer and economic development in the complex sphere of agriculture and natural resources. We find that agricultural producers, for example, are more computer literate than many of the small manufacturers, and they are quick to recognize and adopt new production technologies. But it is clear that the economic and social future of postagrarian rural America is now critically dependent on accessing a broader spectrum of the expertise in the land-grant universities than in the past. Rural economic development—broadly defined—needs to include manufacturing and other sources of off-the-farm employment, new product designs, and more diverse marketing as well as traditional agricultural production and processing activities. In Michigan, the small manufacturing supplier industry, many service companies, and the complex tourism industry are an integral part of the rural economic fabric and must be treated as such.

Much of the scientific foundations and the engineering, business, and organizational skills required to provide leadership in rural economic development exists in the research, education, and extension programs at most land-grant universities. MSU supplements its resources with those of its four sister institutions through TTN. The complementary dimension of the TTN as an interinstitutional network serving small and medium-size industrial firms, and CES with its network of field agents functioning in virtually every county of the state, makes CES/TTN potentially a very powerful joint instrument of economic development, especially in rural America.

What we require is: (1) a philosophical outlook and commitment on the part of the faculty and administrative units to research, education, and extension programs that are driven—at least in part—by the new economic development issues; and (2) coherence in the specific goals, objectives, and priorities among specific initiatives selected to deal effectively with these enormously complex and challenging issues. To this end, MSU has established a number of consortial agreements with sister educational institutions in the state in an attempt to complement the academic resources of participating institutions in addressing regional economic competitiveness and diversification issues. Technical colleges, community colleges, and public and private four-year institutions collectively can provide an impressive array of assistance.

Special Arrangements with the Private Sector

MSU's priority commitment to new start-up firms in fostering economic competitiveness is illustrated by our encouragement of the MSU Foundation to use part of its investment portfolio in setting up a small for-profit biotechnology corporation called Neogen, based on MSU scientific discoveries. Further, we have taken several steps to encourage faculty experts who are so inclined to become involved in the formation of companies that interact with the university. A list of these faculty-based new start-ups is given in Figure 1.9 along with example "wiring diagrams" showing how academic resources are typically linked to the world of commerce through new start-ups. These steps have required changes in university policies and procedures as well as active facilitation.

The results have been very beneficial. In six years, Neogen—for example—has grown from a small office and a part-time president to a thriving company with 80 employees and three subsidiaries. Meridian Instruments, a maker of highly sophisticated instruments for sorting and analyzing cells, has sold 19 instruments valued at a total of $3.5 million to Japan. The instrument has also been awarded a Michigan Product of the Year award, and cell research made possible by it has been featured on CNN-TV and as the cover article in *Science*.

Figure 1.9   Private-sector Examples of Bridges between Business and Academe

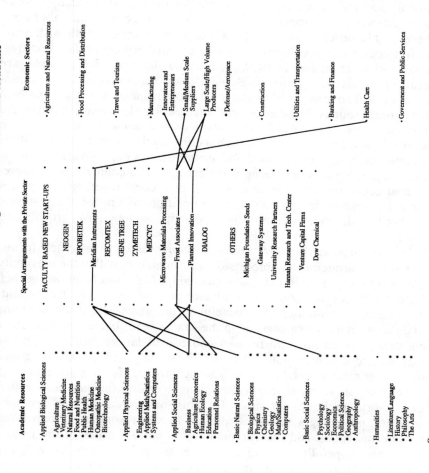

Source: The authors.

Adaptations in Educational Programs and Institutional Policies

In the long-term the strongest link between academe and the economy will be through the stream of graduates and lifelong-education enrollees who ultimately assume roles and responsibilities at various levels of the private and public sectors of the world's economy. Historically science and technology, business and finance, and management and labor have been highly differentiated fields of study and practice, both within the universities and in the corporate world. Seventy percent of the executive-level corporate leaders in the U.S. Fortune 500 firms, for example, are drawn from the pool of graduates with formal training in business administration and no formal training in science or engineering. Only 30 percent have backgrounds in science and engineering. In Japan, exactly the reverse is true. Graduates from highly unrelated academic programs such as business administration and engineering tend to remain highly differentiated in the corporate world and are sometimes even adversarial in the work place. Yet it is increasingly evident that, in a competitive entrepreneurial economy, all of these elements must come together at unprecedented levels of integration and cooperation to ensure the in-house market-based, planned innovation essential to long-term growth and vigor.

Just as corporate organizations and cultures must assume new forms, so too must academic and professional parochialism within the university give way to enlightened catholic scholarship. MSU has been taking such factors into consideration for a number of years in its faculty appointments and in promoting transdisciplinary programs, and it will continue to do so even more aggressively in the years ahead. Indeed, as budgetary constraints force program reviews and consolidations, such integration may offer both opportunities for efficiencies in operation and programs that better meet societal needs and attract stronger funding streams.

The base of faculty support is beginning to emerge that can put together new hybrid programs of study for both undergraduate and graduate students, providing a new generation of human resources to help mainstream innovation and entrepreneurship in various areas of the private sector and hopefully to bring changes in the public sector as well. But corporate America cannot wait for the next generation of students to develop into mature leadership, and no student can or should expect the formal educational process to terminate on graduation. The new hybrid programs of study under development will be designed with lifelong education specifically in mind. That is, these programs will be structured as the first phase of a more comprehensive lifelong educational program to which MSU is making a major and lasting commitment, and to which the graduate can look as an integral part of his or her professional development in the corporate world.

This level of refocusing of the traditional AAU/land-grant institution will require continuing review of university policies and practices to ease present rigidities and develop a long-term recommitment to the mission of fueling and sustaining the nation's economic development in the intensifying international economic, technological, and cultural competition.

Notes

1. All individuals and cultures are consumers, but not all have the opportunity, capability, or motivation to make a net contribution and thus be resources— because of age, health, or cultural opportunity. Central innovation, resource allocation, and tradition can work to move larger or smaller numbers of individuals in both directions between the two categories.

2. Capital is defined as claims on material and human resources.

3. A mature industry is one in which production capacity exceeds market potential.

4. ITI is located near the University of Michigan; MBI is located near Michigan State University; and MMPI has no permanent location at present.

Readings

Pat Choate, Senior Economic Analyst, Center for Occupational Research and Development, and Linger, J.K., TRW, Inc., Arlington, Va. "Preparing for Change." 1987.

Executive Office of the President. *The State of Small Business.* Washington, D.C.: Government Printing Office, 1984.

Milton Stewart, Small Business High Technology Institute. Private communication. 1987.

SRI International, Center for Economic Competitiveness. "Technology and the Michigan Economy: Elements of a Strategy." Summary Report prepared for the Michigan Strategic Fund. October 1987.

2

The Role of the Research University in Creating and Sustaining the U.S. Technopolis

David V. Gibson and Raymond W. Smilor

This chapter takes the position that a high-quality research university or institute is a necessary but not sufficient condition for the creation and maintenance of economic development in the technopolis.[1] As affordable and reliable transportation facilities (e.g., rail and water) and an affordable labor pool were crucial determinants of smokestack industry foundings and growth, a highly educated, professional work force and an affordable high quality of life are key to the creation and maintenance of high-tech industry. The research university educates and stimulates the scientists and engineers necessary for the research activities on leading-edge technologies associated with the technopolis.[2] There also is the key role that the university plays in providing the talent and professionally competent and managerially adept people to combine scientific research and invention with the practical applications of technology. And the university is an important source of liberal arts and cultural activities that underpin the quality-of-life factors necessary to sustain technopoleis. However, as necessary as the research university is to the creation and maintenance of a technopolis, public and private sectors representing a range of institutions are also important to the high-tech development of a region.

We use a conceptual framework called the "Technopolis Wheel" (Figure 2.1) to describe the complexity of technology development and economic growth in a technopolis (Smilor, Kozmetsky, and Gibson 1988). The wheel emphasizes seven major segments in the institutional makeup of a technopolis: the research university, large technology companies, small technology companies, state government, local government, federal government, and support groups. This chapter focuses on the central role

Figure 2.1   The Technopolis Wheel

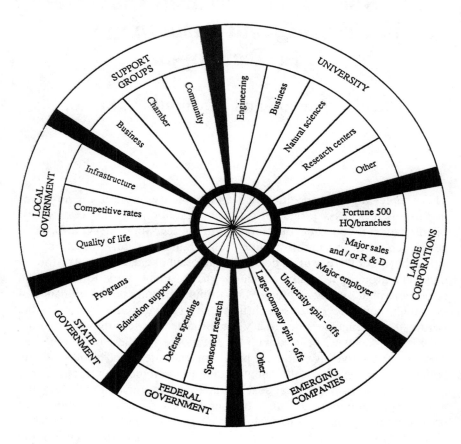

Source: The authors.

of the research university in creating and sustaining a technopolis; we investigate examples from a mature technopolis (Silicon Valley, California), a developing technopolis (Austin, Texas), and an emerging technopolis (Phoenix, Arizona).[3]

On the one hand, a great deal of competition takes place between a state's universities, companies, and public- and private-sector entities. On the other hand, cooperation is essential for a technopolis to develop and survive over time. The various segments of the Technopolis Wheel must find ways to cooperate while competing. To realize such cooperation concerning the role of the research university, this chapter emphasizes the importance of networking across the seven segments of the Technopolis Wheel—that is, the ability to link public- and private-sector entities, (some of which have been traditionally adversarial) in order to effect change.

As new kinds of institutional developments among business, government, and academia are beginning to promote economic development and technology diversification (Allen and Levine 1986; Ryans and Shanklin 1986; Reynolds 1987; Merrifield 1987; Sexton and Smilor 1986), a fascinating paradox has emerged—the paradox of competition and cooperation, which Ouchi (1984) elaborates in his description of the M-Form Society:

> The essence of an M-Form society is social integration. An M-Form society represents balance, a balance between the need for government regulation and the need for independent laissez-faire action. A balance between one special interest and another. (Ouchi 1984, 226)

Case Studies

Silicon Valley

Just as Manchester, the Saar Valley, and Pittsburgh were once centers of the industrial society, today's U.S. technology-driven society has its heartland in Silicon Valley, a 30-by-10-mile strip between San Francisco and San Jose, California. Rogers and Larsen in their book *Silicon Valley Fever: Growth of High-technology Culture* (1984) identified the area as the nation's ninth largest manufacturing center, with sales of more than $40 billion annually. About 40,000 new jobs were created in the area each year in the early 1980s, and the area's economy was among the fastest-growing and wealthiest in the United States.

According to Larsen and Rogers (1988), Silicon Valley is now in its second, more mature phase. The first period of growth, from 1960 to 1980, is over. The attention generated by the expansion of larger centralized corporations is shifting to smaller decentralized companies. Whereas Silicon Valley's earlier years addressed semiconductor and computer manufacturing and its adolescent years saw semiconductor chips and computers being used by other industries, by the late 1980s new subindustries were emerging as a result of innovative microelectronics applications such as medical electronics, communication systems, automotive electronics, telecommunications, and biotechnology. Instead of two main industries—semiconductors, and computers—Silicon Valley now serves as headquarters for a broad base of microelectronics applications industries.

Also, according to Larsen and Rogers (1988), during the 1980s a basic value change occurred in Silicon Valley—from strict competition to certain forms of collaboration. These authors note that much competition continues to exist as firms in the same industry seek to outperform each other. But these competing companies are also likely to share technology licensing agreements, to be costockholders in a university-based R&D consortium (like the MCC [Microelectronics and Computer Technology Corporation] and Sematech [a consortium of major U.S. semiconductor producers and users], which are located in Austin, Texas; or the Center for Integrated Systems at Stanford University), and to join in other relationships in which neither party has complete control over the others. As Larsen and Rogers (1988, 114) state, "Much of the stress on collaboration in the microelectronics industry in the 1980s is due to the threat of international competition, especially from Japan. Silicon Valley is beginning to display certain characteristics of the M-Form Society, described by Ouchi (1984)."

Austin, Texas

The early 1980s were special years for Texans because of the state's approaching sesquicentennial in 1986, and centennial celebrations at the state's two flagship universities: the University of Texas at Austin, and Texas A&M University. The development of the Austin technopolis reached a crescendo in 1983 when the Microelectronics and Computer Technology Corporation (MCC) chose Austin as its headquarters after a major and very public site selection process among some of the most visible high-tech centers in the United States. Austin made headlines in the *New York Times,* the *Wall Street Journal* and the world press as the next great Silicon Valley. Nicknamed "Silicon Prairie," "Silicon Gulch," and "Silicon Hills," the area experienced an unprecedented wave of enthusiasm along with the perception that it had suddenly become a major technology center.

In 1984 the dramatic and unexpected plunge in oil prices, coupled with declining farm and beef prices, caused a general economic decline in Texas: A state that had previously enjoyed a budget surplus and no corporate or personal income taxes now faced budget deficits. The development of Austin experienced a series of problems revolving around a general economic recession in the state, cutbacks in higher education funding, changes in local governmental attitudes, a speculative development cycle that ended in a plethora of foreclosures and bankruptcies, and a general loss of direction.

By early 1988 the effects of an economic recession were still quite apparent in Texas and in Austin. However, the state had begun to increase funding for higher education as well as provide other research support such as the Advanced Technology and Research Program, which was funded to the tune of $60 million by the 70th Texas Legislature with the expressed purpose of supporting economic development in Texas by: (1) attracting the best researchers and students to Texas, and (2) expanding the state's existing technology base. And in early 1988, after a national competition, the main players in the U.S. semiconductor industry chose to locate the industry's new research consortium (SEMATECH) of 13 member companies in Austin.

Phoenix, Arizona

As of 1987, electronics, computers, and aerospace made up the largest portion of manufacturing employment in Arizona—which gave the state a much larger share of high-technology employment than the national average. Five of every ten manufacturing jobs in Arizona are in high technology. Fourteen large companies account for more than three-fourths of the high-technology work force. Almost 80,000 individuals are employed in high technology. Motorola with about 22,000 employees is the state's and also Phoenix's largest high-tech employer (Wigand 1988). With nearly three million people, the state's 39 percent growth from 1973 to 1983 placed it fourth in the nation in rate of population growth. Nationally, the Phoenix metropolitan area ranked second in population growth during 1986–87. Almost 70 percent of the Arizona population is located in the Phoenix metropolitan area. Phoenix proper is the nation's tenth largest city.

According to Wigand (1988), a major key to Arizona's success is the state's employment mix. The goods-producing sector (manufacturing, mining, construction) accounts for about 23 percent of total employment. Manufacturing accounts for almost 14 percent of employment, with half of the manufacturing jobs being in high technology. This compares favorably to the national average of 12 percent of manufacturing employment being related to high technology. Even within Arizona's high-technology component, there is a broad diversity among computers, components, and aerospace. There are cycles associated with the forces of supply and demand that make the state's economy and work force vulnerable to sharp fluctuations. High-technology companies are generally concentrated in certain areas, affecting both the local economy, the employment level, and the appearance of these industrial enclaves (Wigand 1988; Leinberger 1984).

In a 1987 survey of 150 U.S. cities that was based on the number of jobs generated, business start-ups, and young companies enjoying high growth rates, Austin, Texas, was rated first; Orlando, Florida, second; and Phoenix, Arizona, third (Noyes 1987).

The Centrality of the University Segment

In each of the technopoleis discussed in this chapter, the research university has played a key role in the development and maintenance of an area as a technopolis. Stanford University has played a key role in the development of Silicon Valley, and the university continues to contribute to the new subindustries emerging in the area (Rogers and Larsen 1984; Larsen and Rogers 1988). Technology commercialization in the emerging technopolis of Austin, Texas, is underpinned by the University of Texas at Austin (Smilor, Kozmetsky, and Gibson 1988), and Arizona State University is considered key to high-tech development in the Phoenix area (Wigand 1988).

Dependent as they are on technological innovation and a highly educated work force, high-technology companies choose to locate in areas where there is strong research-and-development activity. Much research is conducted by corporations, but the importance of a major research university is well documented (Dempsey 1985; Rogers and Larsen 1984; Botkin 1988; Abetti, LeMaistre, and Wacholder 1988; Ryans and Shanklin 1988). The research university plays a key role in the fostering of research-and-development activities, the attraction of key scholars and talented graduate students, the spin-off of new companies, and the attraction of major technology-based firms; it serves as a magnet for federal and private-sector funding and as a general source of ideas, employees, and consultants for high technology as well as infrastructure companies (Sexton and Smilor 1986; Doutriaux 1987; Smilor, Kozmetsky, and Gibson 1988).

Universities also team with developers, or become developers themselves, in undertaking projects to provide industrial or commercial space and incubator facilities. Some universities have established affiliates directly or by joint venture to conduct research and to provide specialized services to industry. These may have the effect of accelerating innovation while reducing the cost to companies of supporting the research program. It also creates revenues and develops properties—such as research parks—adjacent to the universities (Wigand 1988; Abetti, LeMaistre, and Wacholder 1988).

Rogers and Larsen (1984) cite the resource of Stanford University—and specifically its visionary vice-president, Frederick Terman—as critical to the beginning of Silicon Valley during the 1940s and early 1950s. Many of the early Silicon Valley engineers were Stanford graduates who wanted to remain in the Bay Area, even though in the mid-1900s there was a feeling of inferiority to the big East Coast electronic firms like RCA. Leede Forest (of the Federal Telegraph Company), inventor of the amplifying quality of the vacuum tube, and William Hewlett's and David Packard's entrepreneurial work on the variable-frequency oscillator are prominent examples of the electronics pioneers in the Silicon Valley area who were either educated at Stanford or had university connections.

Rogers and Larsen (1984) consider Terman's conception of a university-affiliated research park his most important contribution to Silicon Valley and to Stanford. In 1951 the idea of a university-affiliated industrial park was a revolutionary idea, so Terman was a visionary. His belief in the value of close university–industry ties led him to suggest leasing a large section of university-owned land to high-technology companies. Hewlett-Packard and Varian Associates were among the first tenants. By 1955 seven companies were in the park; by 1984 there were 90 tenants and 25,000 employees. Stanford Industrial Park has served as a model for scores of other high-tech parks in the United States and abroad; and it did a great deal for Stanford University, such as providing funds with which to hire top professors. In short, Silicon Valley helped establish Stanford University as a world-class, preeminent university. On the other hand, the rise of Stanford's prominence as an internationally recognized research university facilitated the takeoff of Silicon Valley's microelectronics industry and the other high-tech industries—such as biotechnology and telecommunications—that were to follow.

As of 1987 there were more than 300 research parks in the world—more than half of which were located in the United States (General Accounting Office, 1983). Ideally, a park should bring university researchers together with their counterparts in industry, integrating the results produced by both parties. This approach encourages more university research and places the university on the cutting edge of new technological developments (Wigand 1988; Abetti, LeMaistre, and Wacholder 1988).

Smilor, Kozmetsky, and Gibson (1988) contend that the nucleus in the development of the Austin technopolis is the University of Texas at Austin. Indeed, these authors state that, if such a major research university were not in place and had not attained an acceptable level of overall excellence, then the Austin area could not have developed as a technopolis. There would be little or no research-and-development funding, no magnet for the attraction and retention of large technology-based companies, and no base for the development of small technology companies. These authors list three factors as especially important in the development and measurement of a technopolis. The research university plays a key—indeed necessary—role in each, as follows:

1. The achievement of scientific preeminence. A technopolis must earn national and international recognition for the quality of its scientific capabilities and technological prowess. This may be determined by a variety of factors, including R&D contracts and grants; chairs, professorships, and fellowships in universities; membership of faculty and researchers in eminent organizations such as the National Academy of Sciences and the National Academy of Engineering; number of Nobel laureates; and quality of students as measured by the number of National Merit scholars. In addition, scientific and technological preeminence may be measured through newer institutional relationships such as industrial R&D consortia, academic and business collaboration, and research and engineering centers of excellence.

2. Development and maintenance of new technologies for emerging industries. A technopolis must promote the development of new industries based on advancing cutting-edge technology. These industries provide the basis for competitive companies in a global economy and the foundation for economic growth. They may be in the areas of biotechnology, artificial intelligence, new materials, and

advanced information and communication technologies. This factor may be measured through the development of R&D consortia, the commercialization of university intellectual property, and new types of academic–business–government collaboration.

3. Attraction of major technology companies and the creation of home-grown technology companies. A technopolis must impact economic development and technological diversification. This may be deter-mined by the range and type of major technology-based companies attracted to the area, by the ability of the area to encourage and promote the development of homegrown technology-based compa-nies, and by the creation of jobs related to technologically based enterprises.

In the case of the University of Texas at Austin, between 1977 and 1986 there was a substantial increase in the annual total amount of contracts and grants (federal and nonfederal) awarded to the university (Figure 2.2). Smilor, Kozmetsky, and Gibson (1988) attribute much of this in-crease to the UT Endowed Centennial program for chairs, professorships, and fellowships. In other words, centennial endowments made a signifi-cant difference in attracting researchers who, in turn, attract additional research funds.

The number of endowed fellowships, lectureships, professorships and chairs in the University of Texas at Austin increased significantly since 1981. From 1981 through 1986, fellowships and lectureships in business increased from 2 to 68, in engineering from 0 to 67, and in natural sciences from 2 to 50. From 1981 through 1986 professorships in business in-creased from 20 to 71. Thirty-two of these 71 professorships were filled by the end of 1986. Professorships in natural sciences increased from 12 to 75. Forty-three of these 75 professorships were filled by the end of 1986.

Figure 2.3 indicates the cumulative total of endowed chairs in business, engineering, and natural sciences from 1981 through 1986. Chairs in business increased from 3 to 20. Twelve of the 20 chairs were filled by the end of 1986. Chairs in engineering increased from 7 to 34. Twenty of the 34 chairs were filled by the end of 1986. Chairs in natural sciences increased from 3 to 36. Thirteen of the 36 were filled by the end of 1986. While obtaining the financial resources to fund fellowships, professor-ships, and endowed chairs is a significant challenge, recruiting the quali-fied talent is perhaps more difficult. There is intense national and some-

Figure 2.2   Annual Total Amount of Contracts and Grants Awarded to UT–Austin, 1977–86

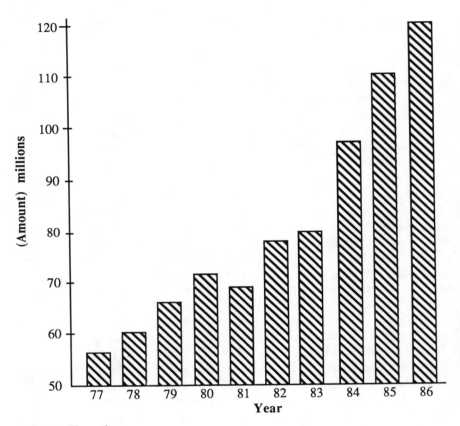

Source: The authors.

times international competition to hire the best professors and to recruit the best students. Of both, there is a limited supply.

In addition, the University of Texas at Austin established major organized research units in the College of Engineering and College of Natural Sciences. Table 2.1 shows 18 research centers in the College of Engineering with a total funding in 1986 of $28,916,099. Table 2.2 shows 32 research centers in the College of Natural Sciences with a total funding in 1986 of $21,354,719. Many of these research units are in emerging, cutting-edge technological areas.

In early 1989, Texas's first technology-targeted business incubator went

Figure 2.3   Cumulative Total of Endowed Chairs to UT–Austin, 1981–86

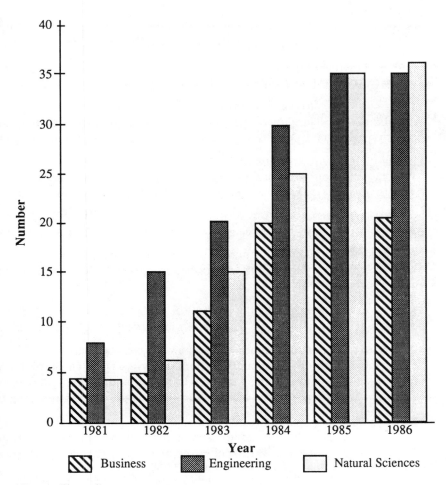

Source: The authors.

into operation. The new venture—the Austin Technology Incubator (ATI)—is designed to stimulate economic development throughout Texas. It is a cooperative venture of the Graduate School of Business and the IC² Institute at the University of Texas at Austin, the City of Austin, the Austin Chamber of Commerce, and private enterprise. The ATI is designed to take advantage of fast-breaking technological discoveries in such areas as new materials, biotechnology, and computer-integrated

Table 2.1   Organized Research Units in Engineering at UT–Austin, January 1987

	Funding Levels
Aeronautical Research Center	$ 415, 491
Texas Institute for Computational Mechanics	415,018
Computer and Vision Research Center	399,487
Construction Industry Institute	1,425,519
Center for Earth Sciences and Engineering	253,266
Electrical Engineering Research Lab	941,150
Center for Electromechanics	11,096,384
Electronics Research Center	699,255
Center for Fusion Engineering	460,081
Geotechnical Engineering Center	461,975
Center for Materials Science and Engineering	1,709,255
Microelectronics Research Center	2,041,240
Center for Enhanced Oil and Gas Recovery Research	691,355
Center for Polymer Research	1,356,817
Center for Space Research and Applications	1,558,102
Phil M. Ferguson Structural Engineering Laboratory	859,081
Center for Transportation Research	3,224,539
Center for Research in Water Resources	908,084
Total	**$ 28,916,099**

Source: The authors.

manufacturing by pooling the resources of the university and various high-tech manufacturing companies and services until the start-up companies have sufficient financial strength to operate independently. Furthermore, the incubator acts as a real-life laboratory for UT faculty and students.

In a study by Larsen, Wigand, and Rogers (1987), more than 70 percent of the respondents within the Phoenix microelectronics industry indicated that the presence or nearness of Arizona State University (ASU) ranked among the top three reasons to locate in the area. ASU is the nation's sixth largest university, with almost 44,000 students. The engineering program at ASU began in 1956 under the guidance of Dean Lee Thomp-

Table 2.2   Organized Research Units in Natural Sciences at UT–Austin, January 1987

	Funding Level
Applied Microbiology, Center for	751,158
Artificial Intelligence Laboratory	1,422,556
Biomedical Research, Institute for	515,489
Brackenridge Field Laboratory	204,422
The Field Station	*
Cell Research Institute	422,686
Central Hybridoma Facility	*
Clayton Foundation Biochemcial Institute	1,154,731
Institute for Computing Science and Computer Application	1,637,166
Culture Collection of Algae	153,005
Developmental Biology, Center for	520,446
Electrochemistry, Laboratory of	551,852
Fast Kinetics Research, Center for	727,458
Fusion Research Center	5,507,251
Fusion Studies, Institute for	2,500,287
Genetics Institute	1,293,788
Ilya Prigogine Center for Studies in Statistical Mechanics	400,311
Materials Chemistry,Center for	*
Nonlinear Dynamics, Center for	424,378
Numerical Analysis, Center for	186,719
Particle Theory, Center for	225,094
Plant Resources Center	53,681
Protein Sequencing Facility	*
Radiocarbon Laboratory	86,896
Reproductive Biology, Institute for	520,323
Relativity, Center for	279,081
Research Institute-Weinberg	*
Research Instruments Laboratory	75,277
Statistical Sciences, Center for	149,281
Structural Studies, Center for	300,297
Theoretical Chemistry, Institute for	863,617
Theoretical Physics	252,603
Vertebrate Paleontology Laboratory	174,866
Total	**$ 21,354,719**

Source: The authors.
*Funding levels for some units are not known.

son, who emphasized the development of undergraduate and graduate programs. The College of Engineering and Applied Sciences conducted research as well, with sponsored research funding averaging about $1 million per year (Beakley, Backus, and Kelly 1985).

In 1979 under Dean C. Roland Haden, the decision was made to review the status of the College of Engineering and Applied Sciences, partially in response to the growing high-technology industrial base forming in the Phoenix area (Wigand 1988). A 50-member Advisory Council of Engineering was organized and comprised leaders from high-technology industry in Arizona, representatives of state government, and ASU engineering faculty. The initial goals of this council were to evaluate the engineering program at ASU and to develop a strategy bringing the college up to national standards. The eventual goal was to make the College of Engineering and Applied Sciences at ASU one of the top schools in the country in graduate studies, and a top research institution (Wigand 1988).

To support the Engineering Excellence Program, a research component composed of four research centers was developed at ASU: the Center for Advanced Research in Transportation, the Center for Solid State Electronics, the Center for Energy Systems, and the Center for Automated Engineering and Robotics. Together these academic and research units emphasize six content areas: solid-state electronics, computers/computer science, computer-aided processes, energy, thermoscience, and transportation (Wigand 1988).

The major strides accomplished by ASU's College of Engineering and Applied Sciences in a five-year time span are exemplified by the growth of the faculty, students, and physical plant. Between 1979 and 1984, there were added 65 new faculty lines (a 59-percent increase) and 52 full-time equivalent new graduate assistant lines (a 33-percent increase) (Beakley, Backus, and Kelly 1985). Sponsored research jumped from just over $1 million in 1979 to approximately $9.5 million in 1984 (an 864-percent increase). The undergraduate population increased from 2,547 in 1979 to 3,351 in 1984 (a 32-percent increase), and graduate enrollments increased from 712 students in 1979 to 977 in 1984 (a 37-percent increase). Finally, the Engineering Excellence Program has been enhanced with a 120,000-square-foot Engineering Research Center that includes a 4,000-square-foot class 100 clean room, portions of which are class 10. According to a 1983 National Academy of Sciences study, ASU's College of Engineering and Applied Sciences is currently in the top 20 engineering programs in the country.

The growing bond between the restructured engineering program at ASU and industry is seen in the development and implementation of a strong continuing-education component. In 1980 a Center for Professional Development was established as a part of the Engineering Excellence Program. The goal of this center is to meet the increasing demand by engineering and applied science professionals for continuous updating and maintenance of their technical competency and skill. From its inception to 1984, more than 160 short courses and institutes were held, and were attended by more than 5,000 professionals (Wigand 1988).

ASU's Interactive Instructional Television Program (IITP) began broadcasting courses to off-university sites in 1981. Computer science and engineering courses are directed to the high-technology companies located in the Phoenix area. As of 1986, there were 15 participating sites. Through a sophisticated teleconferencing system, students at the remote sites are able to interact with the faculty member giving the lecture and with students at other remote locations. This method of providing graduate-level courses and special seminars was initiated at the request of local industry so that their employees could receive graduate degrees.

Consistent with its emphasis on high-technology research, in 1982 the proposal was made that ASU establish a research park to further university–industry cooperation. Development of this concept involved a number of factors. Perhaps most important was the pressure placed on ASU by industry leaders to improve the university's engineering program (Wigand 1988). The feasibility of the park was studied by a committee composed of university and business representatives. In 1983 the ASU Board of Regents approved the proposed park, and in the same year the first executive director was hired. In 1984 the Board of Regents approved the master plan and authorized the creation of a separate seven-member nonprofit corporation: Price-Elliot Research Park, Incorporated. This corporation has as its mission the design, development, marketing, and administration of the park on behalf of ASU for 99 years.

Most academic and business leaders in the Phoenix area believe that the park is a very positive asset for ASU (Larsen, Wigand, and Rogers 1987). As of 1987 two buildings had been constructed, and two additional ones—including the International Microelectronics Innovation Center (IMIC)—had been started. The center had its first tenant: VLSI Technology Incorporated from San Jose, California. Several other tenants had signed leases, including four incubator and start-up firms.

Government Segments and the Research University

Federal, state, and local governments play vital roles in the development of a technopolis. The federal government impacts a state's economic vitality in two key ways: (1) through the development and operation of military installations; and (2) through federal funding for research-and-development activities.

In Texas, Bergstrom Air Force Base—established in 1942—has provided fundamental economic stimulation to the Austin region through its employment of 1,000 civilian and 6,000 military personnel, with an annual payroll of about $167 million (U.S. Government documents). An example of more direct government stimulation to the emerging Austin technopolis is Balcones Research Park, which was created in the early 1940s when the federal government ceded the land to the University of Texas and funded research in strategic resources to support the war effort.

In Texas, as is the case in other regions in the United States, state government is responsible for the major portion of funding for the budgets of public universities. The University of Texas at Austin has benefited tremendously from its Permanent University Fund (PUF), with a 1987 book value at $2.6 billion. This public endowment has been crucial to development of the teaching and research excellence at UT as well as in permitting the acquisition of modern facilities and laboratories. The PUF alone, however, has proven to be insufficient for providing the resources necessary to develop a world-class university.

For example, in 1984 shortly after the MCC decided to locate in Austin and while oil prices were still about $30 a barrel and state revenues increased by $5.4 billion or 17 percent over the previous year, Texas decreased appropriations for higher education by 3 percent. During this time period—despite UT's phenomenal growth in endowed chairs, professorships, lectureships and fellowships; despite the location of the MCC in Austin; and despite national and international press extolling the University of Texas at Austin as a center of excellence in education—the lack of sustained state support for higher education sent a mixed message to the best scholars and researchers whom the university was trying to attract (Gibson and Rogers 1988). During 1984–86, Texas universities in general were not competitive with other U.S. universities in terms of faculty salary. As of 1987, the gap lessened, but UT System faculty salaries still trailed averages offered in the ten most populous states (*Statistical Handbook*).

Arizona's early economic development was primarily resource based—in terms of copper, cotton, citrus, and cattle—and largely controlled by forces and institutions beyond the state's boundaries. Wigand (1988) cites Arizona's recent development strategy as formulated in 1983 under Governor Bruce Babbitt, which specifically addresses high-technology industry.

First, this strategy must put a great deal of emphasis on the high-technology future of Arizona. . . . Small business vitalization must be the second major focus . . . ensuring that the optimal economic potential of all areas of the state is recognized and supported. . . . Finally, strategies for the promotion of technological innovation, business development and balanced economic growth are incomplete if they do not directly address the critical need for education and manpower training. (State of Arizona OEPAD 1984)

While Arizona does have a Science and Technology Advisory Board, it has been largely inactive. According to Wigand (1988), the board could provide a worthwhile focus for technology transfer programs if it were revived and appropriately constituted. Furthermore—states Wigand (1988)—Arizona could take a more aggressive role in encouraging the formation of research-and-development consortia in areas where the state has specialized research capability.

Several studies have been conducted by the Arizona Office of Economic Planning and Development (OEPAD)—now the Arizona Department of Commerce—to define the support industry needed by high-technology development. These reports identify opportunities created by high-technology industries, especially for new service and support industry. The office has established a Small Business Development Corporation to assist in funding businesses. By encouraging companies to start up or expand, Arizona not only assists industry already in the state, but also increases the multiplier effect. Thus, high-technology industry may also affect the ''low-tech'' portions of the state's economy as new products and procedures increase industry efficiency and reduce costs (Wigand 1988).

As of 1988, the Arizona Legislature was discussing bipartisan bills that would enable counties to create enterprise zones and encourage industrial development. Areas designated as enterprise zones would receive tax breaks to lure new business and industry—such as a 5-percent exemption from sales tax on prime contractors, a reduced transaction privilege tax, and an income tax credit of $2,500 for each additional employee added to a firm's payroll.

While state government's primary role in creating and sustaining the technopolis is in relation to setting industrial development priorities and the funding of higher education, local government's primary role generally focuses on the quality of life, competitive rate structures for such items as utilities, and infrastructure requirements. ''Quality of life'' can carry many a different meaning for each person given one's perspective and given the subjective attributes of the issues involved. For example, from the developer's perspective a high quality of life would consist of a viable business environment, which is considered essential to the attraction, expansion, and retention of industry. Factors influencing the business environment include taxes and governmental regulations, the availability of appropriate financing, a positive work ethic, and a favorable relationship among business, the community, and the area's educational institutions. A realistic and consistent state development strategy, with an understanding of the area's current development status and plans for the future, also would be considered important (Smilor, Kozmetsky, and Gibson 1988; Wigand 1988; Ryans and Shanklin 1988).

A competitive tax climate is critical to creating and sustaining the technopolis. Tax rates must be realistic to meet both the state's needs and the corporation's ability to pay while remaining competitive in its industry. Income and property taxes, unemployment and industrial accident insurance rates, and the procedures of taxing authorities all must be reasonable or else companies will competitively suffer, go out of business, or be driven from the state. Most corporations are well informed about comparative tax rates and are knowledgeable about regions that are out of line or impose additional taxes or requirements, such as unitary taxes—thus inhibiting economic development (Wigand 1988).

Companies are reluctant to locate or to remain in areas with overly complex, burdensome regulations or permit processes. Problems with regulations add to the cost of doing business, increase the time and effort required for business operations, and add to a company's uncertainty. Examples of regulatory problems include the interpretation and application of environmental regulations, occupational safety and health standards, and municipal standards (Smilor, Kozmetsky, and Gibson 1988; Wigand 1988).

Since the most talented high-technology employees can choose where they want to live and work, a high quality of life is especially important in the creation and maintenance of a technopolis. For example, annual job turnover in Silicon Valley has traditionally been about 30 percent (Rogers and Larsen 1984). In other words, the average employee would have three different jobs in ten years. Such turnover is encouraged by a shortage of qualified, experienced personnel. To counter such rapid turnover, companies offer to their employees such short-term benefits and incentives as stock options, recreational facilities, and training programs (Rogers and Larsen 1984). At another level of analysis, a community's ability to attract workers to the area and to sustain a quality work force is contingent on the area's perceived desirability as a place to live. Physical setting, natural beauty, climate, and educational opportunities are factors that affect such perceptions, along with cultural activities and outdoor recreation possibilities.

Sustaining quality-of-life attributes during times of economic decline as well as growth has proven to be most challenging for both developing and mature technopoleis. Although they admit that it is a factor not easily quantified, Larsen and Rogers (1988) cite the deteriorating quality-of-life factors behind the geographical spread of Silicon Valley microelectronics corporations since the early 1980s. Dense development of industry and housing has led to paralyzing urban congestion. Commuter access to an area's high-tech facilities is essential. Large companies have significant employment levels and operate multiple shifts, requiring a good system of surface transportation. Multilane streets as well as highway and freeway access are required both for workers and for deliveries. Many companies want their facilities located within reasonable drive of a major airport.

As an example of public and private cooperation in planning for quality-of-life factors, the City of Tempe—in which the ASU Research Park is located—is improving the feeder roads surrounding the site. Furthermore, the park offers tennis courts, an FAA-approved heliport, jogging paths, equestrian trails, ramadas and picnic areas, and three lakes that are fed by treated wastewater from a nearby Motorola manufacturing plant. A conference center is proposed with a hotel site including a child-care learning center, a health management facility, a pool, restaurants, and support service.

Wigand (1988) cites the competitive cost of living as important in attracting people to Arizona. The availability and cost of housing in various price ranges is important, as are food, energy, and transportation costs and personal taxes. Larsen and Rogers (1988) cite high land and housing costs as one factor that has influenced large chip-makers to leave Silicon Valley. Such a high-priced quality of life makes it difficult for Silicon Valley to attract workers, especially workers at lower pay levels. As land prices continue to rise, high-tech companies look for other locations with cheaper electrical power and less expensive materials.

However, in spite of these ominous developments, Larsen and Rogers (1988) report that the move of microelectronics manufacturing from Silicon Valley may not be so serious as it seems from the press reports that herald the "end" of Silicon Valley. Among the newcomers to the valley are LSI Logic, Cypress Semiconductor, VLSI Technology, Sierra Semiconductor, and Integrated Device Technology.

Perceptions about sustaining an affordable, high quality of life vary dramatically within any region undergoing rapid economic growth. There is always the possibility that such growth will diminish the very qualities that caused the area to be so attractive to high-technology companies in the first place. This tension between a sustained quality of life and a sustained economic development has been visible throughout the development of technopoleis.

Shortly after the MCC announced its decision to locate in Austin—in 1983—the region experienced a residential and office building boom that was fueled by the popular press and by land speculators. Land and housing prices skyrocketed until the 1984 dramatic and unexpected plunge in oil prices along with declining farm and beef prices. In late 1984, Austin began to experience a series of problems revolving around a general economic recession in the state. The development of the Austin technopolis began to lose momentum. High office vacancy rates, an overbuilt housing market, bank foreclosures, and business and personal bankruptcies have persisted into 1989. On the other hand, accompanying this severe economic downturn have been falling land and housing prices, a more positive view of business growth from city government, and less of a concern with boom-town threats (such as traffic congestion) to Austin's overall quality of life.

Similarly, while Arizona has a commitment to stimulating further growth in its high-technology sector, this growth has not occurred without negative impacts. Wigand (1988) cites studies that have focused on the dissatisfaction with various quality-of-life aspects in the Phoenix area (Price and Hodge 1984; Lindsey 1987). In Phoenix (as well as Silicon Valley and Austin), environmental pollution and hazardous materials associated with the high-technology industry are a concern. While state officials contend that Arizona does have environmental protection laws covering hazardous wastes, air pollution, and the diversion/invasion of ground and surface waters (Stanton 1984), the popular press suggests that the funding and staffing to administer these laws is not adequate.

During the emergence, growth, and maturation of a technopolis, local government tends to move in cycles that favor either the developers or the environmentalists. Adversarial groups may take a variety of organizational forms representing environmental concerns, labor issues, minority viewpoints, and other community interests. When local government supports economic growth, then the development of the technopolis is more likely to increase: That is, company relocation there is facilitated, and obstacles to development diminish. On the other hand, when local government believes that the quality of life is diminishing, then the development of the technopolis is inhibited: That is, obstacles to development increase (such as high utility rates or slower permitting procedures). The issues become quite complex because quality of life and economic development are two sides of the same coin; each has a vital impact on the other. While environmentalists and developers may disagree on what makes for sensible environmental/developmental policy, most agree that overall quality of life suffers when the people who inhabit the community are out of work and cannot afford to pay the costs associated with infrastructure development, housing, or such features as expanded transportation facilities, parkland and recreational opportunities, cultural activities, and the arts.

Support Groups Segment (Infrastructure) and the Research University

One central reason cited by the MCC and SEMATECH for their locating in Austin was the availability of high-quality graduate students in the fields of computer science and electrical engineering. However, representatives of these two research consortia as well as other high-tech firms in the Austin area also mention the importance of other university-trained talent at the graduate and undergraduate levels as well as the graduates of other local educational facilities, such as community colleges and area public schools.

The availability of a quality work force in a variety of disciplines is basic to the development of high-technology industry. Once a technopolis begins to emerge, a kind of momentum takes hold. High-tech companies choose to locate in an area with a concentration of similar companies, to take advantage of the available labor force and infrastructure services. This clustering produces an economy of scale in training and creates opportunities for support industries devoted to labor-force preparation. For example, few other places in the world can match the pool of experienced, specialized high-tech brainpower of Silicon Valley. The area's high-tech companies depend on individuals who design semiconductor clean rooms, tool delicate fixtures, and design innovative products. Companies that need access to such expertise have little choice but to locate where these intellectual resources are concentrated, creating a further agglomeration of microelectronics firms (Larsen and Rogers 1988).

Business-based groups relate to the emergence of specific components for high-technology support in the practices of Big Eight accounting firms, law firms, major banks, and other companies. These components provide a source of expertise—even when embryonic—and a reference source for those founding and/or running technology-based enterprises. For example, high-tech infrastructure requires a significant capital investment as well as a commitment to fund ongoing operational and maintenance costs. From lending policies that reflect an understanding of industry conditions and practices to venture capital availability, the financial community must be responsive if a technopolis is to grow and mature.

The growth of venture capital (Wetzel 1986 and 1987; Brophy 1986; Robinson 1987; Timmons and Bygrave 1986) provides a good example of the importance of business-based groups to the development of a technopolis. According to Rogers and Larsen (1984), without an availability of venture capital, the proliferation of semiconductor firms in the 1960s and early 1970s could not have occurred in Silicon Valley. More than one-third of the nation's venture capital companies have an office in Silicon Valley.

Venture capital firms serve as intermediaries between investors looking for high returns on their money and entrepreneurs in search of needed capital for their start-ups. Venture capital firms invest their money largely on the basis of the potential value of an entrepreneur's idea—a collateral that conventional bankers consider worthless. Entrepreneurs give up a percentage of the ownership of their new company—often a significant amount—in exchange for acquiring capital.

In 1980 Austin had virtually no venture capital money. However, by 1986 the city had approximately $80 million managed by five firms. The growth was due primarily to two factors—one external, and the other internal (Kozmetsky, Gill, and Smilor 1986). Externally, changes in federal tax laws in 1979, 1981, and 1986 pertaining to capital gains encouraged more investment in venture capital pools. Internally, the perception of Austin as an emerging technology center encouraged the development of homegrown pools. The source of the venture capital included a few individuals knowledgeable about the venture capital process, as well as the major commercial banks in the area. While funds in these pools increased, most venture capital investments continued to be made *outside* the state of Texas. Venture capitalists in Austin, while wanting a local window on technology and company development, did not see enough good deals—that is, fast-growth company potentials—in the region (Kozmetsky, Gill, and Smilor 1986).

In terms of fundamental operational and maintenance costs of an emerging technopolis, Arizona is severely limited in its ability to finance the development of public facilities—especially in advance of their need—because the state's constitution places stringent limits on state debt, and subsequent limitations have been placed on municipal debt and spending (Wigand 1988). These limitations make it difficult to construct public facilities until after the population they serve is in place. The resulting inconvenience and increased cost is grudgingly accepted, since the new-comers are paying their share of the needed facilities. However, in the rush to keep up with growth, adequate maintenance programs for public facilities have been overlooked. As Arizona's development matures, the state will meet the same maintenance, upgrading, and replacement issues faced by Eastern states decades ago, and Arizona will pay the price for deferring these expenditures (Wigand 1988).

High-technology industry has other significant infrastructure require-ments, such as water and sewer systems. For example, large companies may use 3 million gallons of water in their daily industrial processes. Arizona is perceived by many as a desert state with limited groundwater overdrafting. Others claim that there is no water-related reason not to locate high-tech industry in Arizona (Steiner 1985). Moderates argue that, although water resources always need careful management, Arizona's water resources are adequate to meet foreseeable municipal and industrial needs (McNulty and Woodard 1984). In Arizona, after industrial use, most of the water returns to the community's sewer system and treatment plant. Due to cost and treatment-plant capacity considerations, many of the larger new developments in the Phoenix area now are building systems that treat sewage on-site and use the effluent for landscape and golf-course irrigation (Wigand 1988).

High-technology companies also require a large supply of reliable, high-quality electric power, and any interruption such as voltage fluctua-tions—no matter how brief—can cause the loss of millions of dollars worth of work-in-progress. While electrical-energy distribution capabili-ties vary in Arizona, Wigand (1988) reports that they are generally adequate to support considerable load growth. Power costs in Arizona are in the midrange of costs nationally, and Arizona's electric utilities are making an effort to attract "high load factor" customers to generate capacity.

The Private Sector and the Research University

While public attention on Silicon Valley usually concentrates on the 54 electronics firms with more than 1,000 employees (e.g., Hewlett-Packard, Intel, and Apple Computer), Rogers and Larsen (1984) note that more than two-thirds of Silicon Valley firms have fewer than ten employees, and 85 percent have fewer than 50. Companies are constantly starting up, growing, merging, being acquired, or fading away—making it difficult to know exactly how many firms exist at any one given time.

Taking the most accurate census they could, Rogers and Larsen (1984) identified 3,100 electronics manufacturing firms in Silicon Valley. In addition, they emphasized the importance of supporting companies— such as firms engaged in marketing, advertising, research and development, consulting, training, venture capital, legal, and other services— which brought the total number of firms in the Silicon Valley electronics industry to about 6,000. The researchers noted that another 2,000 companies engage in business in nonelectronics high-technology fields such as chemicals, pharmaceuticals, and biotechnology. According to their final count, the total number of high-tech firms in Silicon Valley was about 8,000 in 1983.

An even more difficult way to measure the growth of high-technology company development in a technopolis is to track employment and high-technology incorporations over time. Since the mid-1950s—when Motorola established its microelectronics facilities in Phoenix—high-technology manufacturing (computers, electronic components, aerospace, communications, and scientific instrumentation) has accounted for almost 50 percent of all manufacturing jobs in Arizona, compared to a 15-percent national average (Wigand 1988). In 1987 the state had 313 high-technology firms—up 33 percent from 225 firms in 1975—including an electronics firm founded by Hopi Indians (Webster 1987). The value of shipments has exceeded $5 billion annually (State of Arizona OEPAD 1984). Some 85,000 workers are currently employed in high-technology companies in Arizona—a growth of more than double the 42,000 employed in 1975.

Arizona has experienced a high-technology growth rate of about 6 percent annually over the past few years, while U.S. high-technology employment has grown by 2.4 percent annually (Wigand 1988). High-technology employment represents 7 percent of total employment in Arizona, and 3 percent nationally (Price and Hodge 1984). The state's 14 largest firms employ 78 percent of Arizona's high-technology work force—indicating, perhaps, less entrepreneurial activity than is associated with the small firms that exist in Silicon Valley. Metropolitan Phoenix accounts for 79 percent of Arizona's high-technology jobs, and Tucson accounts for 20 percent. Wigand (1988) reports that there are some indicators that a high-technology Phoenix–Tucson corridor may come about. Several high-technology firms have located their facilities between these two regions.

The source of most high-technology growth in Arizona is the expansion of existing companies to new facilities or the relocation of national corporations to Arizona. Companies already in Arizona usually expand to adjacent sites, especially if their operational needs are met and the labor market is not saturated. Between 1975 and 1979, numerous new firms moved to Arizona—including Digital Equipment, IBM, Intel, GTE, and Gould. Many firms made major improvements and expansions within their existing Arizona facilities—including Honeywell, Garrett, Sperry, and Motorola (the state's largest high-technology employer, with about 22,000 employees). While corporate expansion generally does not receive a great amount of attention, Wigand (1988) emphasizes that it represents an important source of growth and serves as an important indicator of the overall economic stability of an area.

Figure 2.4 shows the incorporation of manufacturing-related technology firms in Austin, Texas. Services-related technology firms are not included. In 1984, growth of these firms leveled off—probably as a result of the general economic recession. For Austin, the location and home-grown development of major technology-based companies had its tentative beginnings in 1955. As shown in the timetable in Figure 2.5, Austin had 33 such major-company relocations or foundings as of 1988. Two four-year clusters are interesting to note: 1965–69 and 1980–86. Major events took place in each of these clusters. During the first, IBM located in Austin; during the second, the MCC located there.

Six of the 33 companies are homegrown, and all six had direct or indirect ties to the University of Texas at Austin. An important way to assess the impact of UT in the development of the Austin technopolis is to consider spin-off companies. Of 103 small and medium-size technology-based companies in existence in Austin in 1986, 53—or 52 percent—indicated a direct or indirect tie originally to the University of Texas at Austin (Figure 2.6). These companies' founders were UT students, graduates, faculty members, and other UT employees. Their tie to the university had enabled many of them to start their own businesses with a contract that originated while they were involved in university research activities. In addition, the ability to continue their relationship in some capacity with the university was an influential factor in their staying in the area (Smilor, Gibson, and Dietrich 1990). These firms demonstrate an important requirement for a technopolis: the ability to generate home-grown or indigenous technology-based companies, which in turn have a direct impact on job creation and economic diversification.

According to Rogers and Larsen (1984), the rate of spin-offs from a parent high-tech company is the key indicator of the rise (and fall) of a technopolis. The general pattern for locating spin-off companies is ag-

Figure 2.4   Cumulative Total of High-tech Manufacturing Companies in Austin, 1945–85

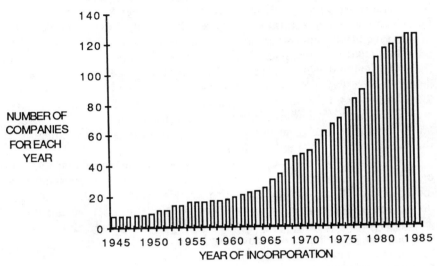

Note: These companies were defined by the following three-digit SIC codes: 283, 348, 357, 364–367, 369, 376, 379, 381–387. A number of studies have incorporated this definition of high-technology products in analysis of high-technology manufacturing. See Massachusetts Division of Employment Security, Job Market Research Division, *High Technology Employment: Massachusetts and Selected States 1975–1979* (Boston: Massachusetts Division of Employment Security, March 1981); Peter Doeringer and Patricia Pannell, "Manpower Strategies for New England's High Technology Sector," paper presented at Conference on Manpower Policy Issues, sponsored by the Commission on Higher Education and the Economy of New England at the Harvard University Graduate School of Business Administration, May 15, 1981; and U.S. Congress, Joint Economic Committee, *Location of High Technology Firms and Regional Economic Development* (Washington, D.C.: Government Printing Office, June 1981).

Source: 1986 Directory of Texas Manufacturers, Bureau of Business Research, Graduate School of Business, University of Texas at Austin.

glomeration. Once a critical mass of companies in a high-technology industry is located in one area, new companies in that industry will concentrate in the area. Such was the case with semiconductor companies in Silicon Valley.

Dr. William Shockley was coinventor of the transistor at Bell Labs in 1947. Shockley moved to Palo Alto, California, in 1955 to found Shockley Transistor Laboratory, which became the direct or indirect source of most of the 80 semiconductor firms that eventually started up in Silicon Valley (Rogers and Larsen 1984). Although his own entrepreneurial venture was short-lived, Shockley nevertheless made a major contribu-

Figure 2.5 Major Company Relocations or Foundings in Austin, 1955–88

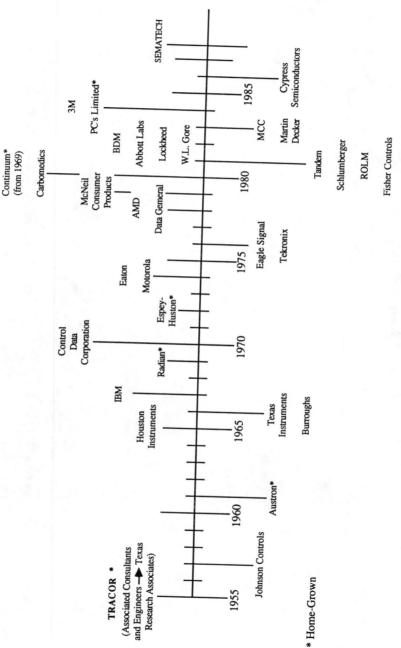

Source: This timeline was developed by the authors on information from the Austin Chamber of Commerce and from survey and interview data, the IC² Institute, Austin, Texas.

Figure 2.6   Small High-tech Firms Founded with UT Connections, 1955–86

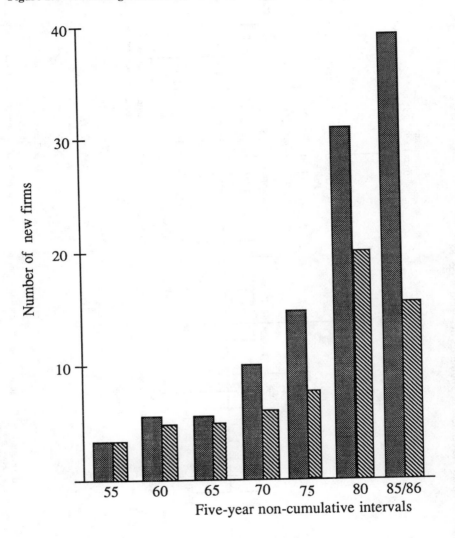

Source: Smilor, Kozmetsky, and Gibson, 1988.

tion to the rise of Silicon Valley by identifying brilliant personnel. Among the bright young scientists whom he recruited to join his company were engineers and physicists on the cutting edge of semiconductor technology—most of them from the East Coast. A select group of these engineers—called the "Shockley Eight"—demonstrated well the entrepreneurial spirit that they learned from Shockley. Within a year, all eight left Shockley Transistor Laboratory to start a semiconductor manufacturing company of their own: Fairchild Semiconductor. The Shockley Eight then became the cadre of leaders for the semiconductor industry that sprouted in Silicon Valley. Fairchild Semiconductor was the spawning ground for scores of spin-offs (Rogers and Larsen 1984).

Silicon Valley's star entrepreneurs receive much attention from the mass media. They serve as much-admired role models, adding further to the entrepreneurial fever in the area. According to Rogers and Larsen (1984), entrepreneurship is best learned by example. When an individual learns of successful Silicon Valley role models like Bill Hewlett and David Packard (Hewlett-Packard), Steve Jobs and Steve Wozniak (Apple Computer), and others, they begin to think, "If he did it, why can't I?" Such role models become a beacon to those who dream of starting their own firms. Once semiconductor firms were established in Silicon Valley, the area became a center for entrepreneurial behavior in the new industries that applied semiconductors to particular uses—microcomputers, telecommunications, and video games, for example.

The Tracor Case

The centrality of the research university to the development of a technopolis and the influence of an entrepreneurial role model to subsequent spin-offs from the parent company can be effectively demonstrated through a case study of Tracor, Inc.—a homegrown company that was, in 1987, the only Fortune 500 company headquartered in Austin. Tracor exemplifies what Kanter (1985) calls a "high innovation company" and what Cooper (1985) calls an "incubator organization."

Frank McBee, the founder of Tracor, earned both bachelor's (1947) and master's (1950) degrees in mechanical engineering at UT after serving as an Army Air Corps engineer from 1943 to 1946. In the late 1940s, McBee became an instructor and then an assistant professor in the UT Department of Mechanical Engineering. In 1950 he became the supervisor of the Mechanical Engineering Department of UT's Defense Research

Laboratory (now called the Applied Research Laboratory) at UT's Balcones Research Park.

In 1955, with $10,000 funding, McBee joined forces with three UT physicists to form Associated Consultants and Engineers, Incorporated—an engineering and consulting firm. Drawing on their UT training and work experience, the four scientists focused their efforts on acoustics research. They were awarded a $5,000 contract for an industrial noise reduction project. The company's name was changed to Texas Research Associates (TRA) in 1957. During the late 1950s, the four scientists taught and did research at UT while working on developing TRA. In 1962 the firm merged with a company called Textran and adopted its present name of Tracor. By this time, McBee had left the University of Texas to devote full time to building the company.

Figure 2.7 pictures how the educated talent from the College of Engineering and the Defense Research Laboratory at the University of Texas at Austin came to form the entrepreneurial venture of Associated Consultants and Engineers in 1955, which eventually led to the establishment of Tracor in 1962. However, even more important is the stream of entrepreneurial talent that came from Tracor after 1965, beginning with the spin-off Pinson Associates, Incorporated. At least 16 companies have spun off Tracor and located themselves in Austin.

Figure 2.8 dramatically shows the job creation impact of Tracor and its spin-offs on the Austin area. A total of about 5,467 persons were employed in these companies as of 1985. Some of these companies have created spin-offs of their own. Radian Corporation, for example, has spun off four companies. Most importantly, neither Tracor, its spin-offs, nor the jobs they created would exist without the University of Texas at Austin, which sowed the seeds of the original venture.

Influencers and the Research University

Rogers and Larsen (1984) describe Silicon Valley as a network of networks. Extensive personal contacts facilitate information exchange so that news about people changing jobs, about new products, or about manufacturing successes and failures all are instantly common knowledge. While each of the institutional segments in the Technopolis Wheel are important to high-technology company development, the ability to link or network the segments is most critical (Birley 1985; Aldrich and Zimmer 1986). Indeed, unless the segments are linked in a synergistic way, the development of the technopolis slows or stops. In Austin, these

Figure 2.7   Development of Tracor and Its Spin-offs, 1947–84

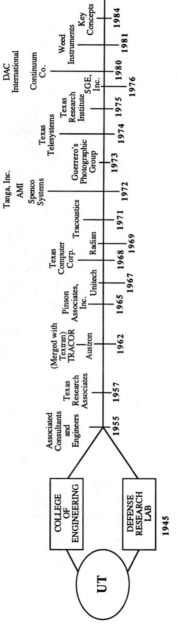

Source: Smilor, Kozmetsky, and Gibson 1988.

Figure 2.8   Job Creation Impact of Tracor and Its Spin-offs

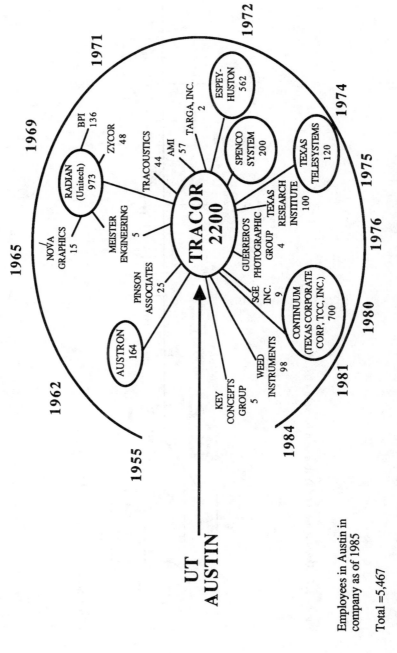

Employees in Austin in
company as of 1985

Total =5,467

Source: This chart was developed by the authors based on survey, interview, and archival data.

segments have been linked by first- and second-level influencers: key individuals who make things happen and who are able to link themselves with other influencers in each of the other segments as well as within each segment.

First-level influencers have a number of criteria in common:

1. They provide leadership in their specific segment because of their recognized success in that segment;

2. They maintain extensive personal and professional links to all or almost all of the other segments;

3. They are highly educated;

4. They move in and out of the other segments with ease;

5. They are perceived to have credibility by influencers in the other segments; and

6. They are visionaries who articulate concepts and objectives for the larger community.[4]

Cross-segment linkage is facilitated by second-level influencers who also represent business, academia, and government as well as local community interests. This level of influencer is more likely to be associated with day-to-day management of the particular projects that are championed by the first-level influencers or visionaries. Within each segment, the second-level influencer/manager interacts with and generally has the confidence of the first-level influencer/visionary. The role and scope of the second-level influencer is to act as a gatekeeper in terms of increasing or decreasing flows of information to first-level influencers. Second-level influencers also have their own linkages to other second-level influencers in the other institutional segments. Working together, first-level and second-level influencers initiate new organizational arrangements to institutionalize the linkage between business, government, and academia.

Influencers seem to coalesce around key events or activities, as described by Gibson and Rogers (1988) in their research on the interstate competition for the MCC. They play a crucial role in conception, initiation, implementation, and coordination of the events or activities. Once an event or action has been successfully managed or achieved, they often help to institutionalize the process so that it can function effectively without them. Influencers play a particularly important networking role through support groups, which can rovide convenient opportunities to interact across all segments of the Technopolis Wheel.

An important role of the research university in a technopolis is to facilitate the recruitment and development of first- and second-level influencers for all segments of the Technopolis Wheel. Based on the present research and the work of others (Rogers and Kincaid 1981; Ouchi 1984; Aldrich and Zimmer 1986), it can be argued that the more extensive and higher the level are the networks across the different segments of the Technopolis Wheel, the more likely it is that cooperative economic (and other) activities will take place at the community and state levels.

Conclusion

The research university plays a pivotal role in creating and sustaining the technopolis. It contributes to:

1. Achieving scientific preeminence;

2. Creating, developing, and maintaining new technologies for emerging industries;

3. Educating and training the required work force and professions for economic development through technology;

4. Attracting large technology companies;

5. Promoting the development of homegrown technologies;

6. Contributing to improved quality of life and culture; and

7. Facilitating the recruitment and development of first- and second-level influences.

Local government can utilize the resources of the research university in creating public policy that will balance quality-of-life issues with company formation and relocation. State government, through its allocations to and commitments toward the university, also affects the development of a technopolis. It plays an important role especially in the areas of making and keeping long-term commitments to fund R&D, faculty salaries, and student support at all levels of educational development. The federal government plays a proactive role for the research university through its allocation of research-and-development moneys, on-site R&D programs, and defense-related activities.

Large technology companies play a catalytic role in the expansion of a technopolis through ties with the research university that include providing adjunct professors, funding research, hiring students, and promoting technology transfer. Large companies are also a source of talent for the development of new companies, which contribute to job creation and an economic base that can support an affordable quality of life. Small technology companies facilitate the growth and maintenance of a technopolis by diversifying and broadening the economic base of the area; contributing to job creation; and spinning off and commercializing technologies developed at the university, other research institutes, and corporations.

The Technopolis Wheel provides a conceptual framework for assessing the relative importance of the research university as well as other institutions and organizations in creating and sustaining the development of a technopolis—as the cases of mature (Silicon Valley), developing (Austin), and emerging (Phoenix) technopoleis demonstrate. By focusing on the interaction among the government, business, and public sectors, these three cases provide a practical perspective on the changing nature of economic development and the importance of new kinds of institutional relationships revolving around the research university. The research university can play a critical role in facilitating the network among individuals and institutions in all segments of the Technopolis Wheel.

Acknowledgments

The authors acknowledge the assistance of the IC^2 Institute staff in helping to prepare the data on the Austin, Texas, technopolis. Special appreciation is extended to Everett M. Rogers (Walter Annenberg Professor at the Annenberg School of Communications, University of Southern California, Los Angeles), Dr. Judith K. Larsen (Data Quest Incorporated, Palo Alto, California), and Rolf T. Wigand (Professor, School of Public Affairs, Arizona State University, Tempe) for their research and writings on Silicon Valley, California, and Phoenix, Arizona.

Notes

1. The term "technopolis" is composed of *techno,* which reflects an emphasis on technology, and *polis,* which is the Greek word for city-state and reflects a balance between the public and private sectors. The plural form of the Greek word *polis* is *poleis.* Therefore we have chosen to use the plural "technopoleis" rather than other possible plural forms such as "technopolises" or "technopoli." See Smilor, Kozmetsky, and Gibson (1988); and Smilor, Gibson, and Kozmetsky (1988).

2. As defined by Rogers and Larsen (1984), high-tech industry is characterized by: (1) highly educated employees, many of whom are scientists and engineers; (2) a rapid rate of technological innovation; (3) a high ratio of research-and-development expenditures to sales (typically about 1:10); and (4) a worldwide market for products. The main high-tech industries are electronics, aerospace, pharmaceuticals, instrumentation, and genetic engineering. Certain subindustries within electronics—like semiconductors and microcomputers—possess a technology that is advancing most rapidly.

3. The description of Silicon Valley is based on the work of E. Rogers and J. Larsen, authors of *Silicon Valley Fever: Growth of High-technology Culture* (1984) and "Silicon Valley: The Rise and Falling Off of Entrepreneurial Fever" (Larsen and Rogers 1988). The tables and graphic presentations on Austin, Texas, are based on interview, survey, and archival data collected by the IC^2 Institute during January–March 1987. Interviews were conducted with respondents who either represented or were knowledgeable of the academic, business, community, and government interests of the region. A telephone survey was used to collect current information on start-up company spin-offs from the University of Texas at Austin and large Austin-based companies. Special attention was given to the case of Tracor, Incorporated—Austin's only homegrown Fortune 500 company. Interview and survey data were checked against archival data whenever possible. The description of Phoenix is based on "High Technology Development in the Phoenix Area: Taming the Desert" by R. Wigand (1988).

4. Examples of such first-level influencers in Texas in 1983–86 include the following: representing state government, Governor Mark White and Mayor Henry Cisneros (San Antonio); representing business, Ross Perot (investor, Dallas), Frank McBee (founder and president of Tracor), and Bobby Ray Inman (president and CEO of the MCC); representing academia, Robert Baldwin (vice-chairman, UT Board of Regents), Hans Mark (UT chancellor), and George Kozmetsky (former dean of UT's College and Graduate School of Business and currently director of the IC^2 Institute).

References

P. A. Abetti, C. W. LeMaistre, and M. H. Wacholder, "The Role of Rensselaer Polytechnic Institute: Technopolis Development in a Mature Industrial Area," in *Creating the Technopolis: Linking Technology Commercialization and Economic Development,* R. W. Smilor, G. Kozmetsky, and D. V. Gibson, eds. (Cambridge, Mass.: Ballinger Publishing, 1988), pp. 125–44.

F. G. Adams and N. J. Glickman, *Modeling the Multiregional Economic System.* (Lexington, Mass.: Lexington Books, 1980).

H. Aldrich and C. Zimmer, "Entrepreneurship through Social Networks," in *The Art and Science of Entrepreneurship,* D. Sexton and R. Smilor, eds. (Cambridge, Mass.: Ballinger Publishing, 1986), pp. 3–23.

D. N. Allen and V. Levine, *Nurturing Advanced Technology Enterprises: Emerging Issues in State and Local Economic Development Policy* (New York: Praeger Publishers, 1986).

G. C. Beakley, C. E. Backus, and R. W. Kelly, "Results of the First 5-year Phase: Excellence in Engineering for the '80s." Report prepared by the Dean's Advisory Council, College of Engineering and Applied Sciences, Arizona State University, Tempe, 1985.

S. Birley, "New Ventures and Employment Growth," *Journal of Business Venturing* 2, 2 (Spring 1987): 155–65.

———, "The Role of Networks in the Entrepreneurial Process," *Journal of Business Venturing* (Winter 1985): 107–17.

J. W. Botkin, "Route 128: Its History and Destiny," in *Creating the Technopolis: Linking Technology Commercialization and Economic Development,* R. W. Smilor, G. Kozmetsky, and D. V. Gibson, eds. (Cambridge, Mass.: Ballinger Publishing, 1988), pp. 117–23.

H. Brooks, L. Liebman, and C. Shelling, *Public–Private Partnership: New Opportunities for Meeting Social Needs* (Cambridge, Mass.: Ballinger Publishing, 1984).

D. J. Brophy, "Venture Capital Research," in *The Art and Science of Entrepreneurship,* D. Sexton and R. Smilor, eds. (Cambridge, Mass.: Ballinger Publishing, 1986), pp. 3–23.

M. Castello, *The Economic Crisis and American Society* (Princeton, N.J.: Princeton University Press, 1980).

A. C. Cooper, "The Role of Incubator Organizations in the Founding of Growth-oriented Firms," *Journal of Business Venturing* (Winter 1985): 75–86.

K. Dempsey, "Hi-tech Race on for Silicon Areas in U.S.," *Plants Sites and Parks* 12 (March/April 1985): 1–22.

W. DeSarbo, I. C. MacMillan, and D. L. Day, "Criteria for Corporate Venturing: Importance Assigned by Managers," *Journal of Business Venturing* 2, 4 (Fall 1987): 329–50.

J. Doutriaux, "Growth Patterns of Academic Entrepreneurial Firms," *Journal of Business Venturing* 2, 4 (Fall 1987): 285–97.

General Accounting Office. *The Federal Role in Fostering University–Industry Cooperation,* PAD-83-22 (Washington, D.C.: GAO, 1983).

J. M. Gibb, *Science Parks and Innovative Centers: Their Economic and Social Impact* (New York: Elsevier, 1985).

D. V. Gibson and E. Rogers, "The MCC Comes to Texas," in *Measuring the Information Society: The Texas Studies,* F. Williams, ed. (New York: Sage, 1988).

A. Glasmeier, "The Japanese Technopolis Program: High Tech Development Strategy or Industrial Policy in Disguise," paper presented to the 29th Annual Conference of the Association of Collegiate Schools of Planning, November 1987.

R. Kanter, "Supporting Innovation and Venture Development in Established Companies," *Journal of Business Venturing* (Winter 1985): 47–60.

G. Kozmetsky, M. D. Gill, Jr., and R. Smilor, *Financing and Managing Fast-growth Companies: The Venture Capital Process* (Lexington, Mass.: Lexington Books, 1986).

J. K. Larsen and E. Rogers, "Silicon Valley: The Rise and Falling Off of Entrepreneurial Fever," in *Creating the Technopolis: Linking Technology Commercialization and Economic Development,* R. W. Smilor, G. Kozmetsky, and D. V. Gibson, eds. (Cambridge, Mass.: Ballinger Publishing, 1988), pp. 99–115.

J. K. Larsen, R. T. Wigand, and E. M. Rogers, "Industry–University Technology Transfer in Microelectronics," report submitted to the National Science Foundation, January 1987.

C. B. Leinberger, "Urban Villages: The Locational Lessons," *Wall Street Journal,* November 13, 1984.

R. Lindsey, "Alarm Raised on Growth of Phoenix," *New York Times,* March 12, 1987, p. 15.

R. Little, "Shoestring Financing of Great Enterprises," *Journal of Business Venturing* 2, 2 (Spring 1987): 1–3.

M. McNulty and G. Woodard, "Arizona Water Issues: Contrasting Economic and Legal Perspectives," *Arizona Review* (1984):1–13.

B. D. Merrifield, "New Business Incubators," *Journal of Business Venturing* 2, 4 (Fall 1987): 277–84.

F. Noyes, "Economic Growth Dependent on Quality of Life, Officials Told." *Arizona Republic,* March 31, 1987, pp. B4, B16.

M. Olson, *The Rise and Decline of Nations: Economic Growth, Stagflation, and Social Rigidities* (New Haven, Conn.: Yale University Press, 1982).

W. G. Ouchi, *The M-Form Society: How American Teamwork Can Recapture the Competitive Edge* (Menlo Park, Calif.: Addison-Wesley Publishing, 1984).

K. Price and C. Hodge, "High-tech: Valley Spinning Shimmering Web of Pure Silicon." *Arizona Republic,* July 23, 1984, p. D1.

R. B. Reich, *The Next American Frontier* (New York: Times Books, 1983).

P. D. Reynolds, "New Firms: Societal Contributions versus Survival Potential," *Journal of Business Venturing* 2, 3 (Summer 1987): 231–46.

R. B. Robinson, Jr., "Emerging Strategies in the Venture Capital Industry," *Journal of Business Venturing* 2, 1 (Winter 1987): 53–77.

E. M. Rogers and D. L. Kincaid, *Communication Networks: Toward a New Paradigm for Research* (New York: Free Press, 1981).

E. M. Rogers and J. K. Larsen, *Silicon Valley Fever: Growth of High-technology Culture* (New York: Basic Books, 1984).

J. K. Ryans and W. L. Shanklin, *Guide to Marketing for Economic Development* (Columbus, Ohio: Publishing Horizons, 1986).

————, "Implementing a High Tech Center Strategy: The Marketing Program," in *Creating the Technopolis: Linking Technology Commercialization and Economic Development,* R. Smilor, G. Kozmetsky, and D. Gibson, eds. (Cambridge, Mass.: Ballinger Publishing, 1988), pp. 209–20.

D. L. Sexton and R. W. Smilor, eds., *The Art and Science of Entrepreneurship* (Cambridge, Mass.: Ballinger Publishing, 1986).

R. W. Smilor, D. V. Gibson, and G. Dietrich, "University Spin-out Companies: Technology Start-Ups from UT-Austin," *Journal of Business Venturing* (1990, forthcoming).

R. W. Smilor, D. V. Gibson, and G. Kozmetsky, "Creating the Technopolis: High Technology Development in Austin, Texas," *Journal of Business Venturing,* no. 4 (1988):49–67.

R. W. Smilor, G. Kozmetsky, and D. V. Gibson, eds., *Creating the Technopolis: Linking Technology Commercialization and Economic Development* (Cambridge, Mass.: Ballinger Publishing, 1988).

S. Stanton, "State's High-tech Growth Requires Strong Pollution Rules, Panel Told," *Arizona Republic,* September 1984, p. B1.

State of Arizona, Department of Economic Security, "Arizona Education—A Statistical Overview," *Arizona Labor Market Newsletter* 8 (June 1984): 21–27.

State of Arizona, Office of Economic Planning and Development (OEPAD), "High Technology in Arizona: A Market Analysis of Suppliers in Arizona and the Southwest," Phoenix, January 1984, p. 4.

Statistical Handbook 1986–1987. University of Texas at Austin, Office of Statistical Studies.

W. E. Steiner, "Water for Municipal and Industrial Growth," State of Arizona, Department of Water Resources, January 1985.

S. Tatsuno, *The Technopolis Strategy* (Reading, Mass.: Addison-Wesley Publishing, 1986).

The Texas Incentive for the MCC, document produced by the Austin, Texas, Task Force for the MCC Site Selection Committee, April 1983.

J. A. Timmons and W. D. Bygrave, "Venture Capital's Role in Financing Innovation for Economic Growth," *Journal of Business Venturing* 1, 2 (Spring 1986): 161–76.

Valley National Bank, *Arizona Progress* 41, 9 (1986):2.

G. Webster, "En-tribe-preneurs: Hopis Use Ancient Skills to Master High-tech Jobs," *Arizona Republic,* March 1987, p. E1.

W. E. Wetzel, Jr., "Informal Risk Capital: Knowns and Unknowns," in *The Art and Science of Entrepreneurship,* D. Sexton and R. Smilor, eds. (Cambridge, Mass.: Ballinger Publishing, 1986), pp. 85–108.

———, "The Informal Venture Capital Market: Aspects of Scale and Market Efficiency," *Journal of Business Venturing* 2, 4 (Fall 1987): 299–313.

R. T. Wigand, "High Technology Development in the Phoenix Area: Taming the Desert," in *Creating the Technopolis: Linking Technology Commercialization and Economic Development,* R. W. Smilor, G. Kozmetsky, and D. V. Gibson, eds. (Cambridge, Mass.: Ballinger Publishing, 1988), pp. 185–202.

3

Socioeconomic Development through Technology Transfer: Tecnopolis Novus Ortus

Umberto Bozzo, David V. Gibson,
Romualdo Sabatelli, and Raymond W. Smilor

In 1969 the University of Bari promoted the seeds of the technopolis concept in Southern Italy with the cooperation of Italian enterprises and local and national government.[1] In previous research, Smilor, Kozmetsky, and Gibson (1988) developed the concept of the Technopolis Wheel in which they identified the interaction of the following seven sectors as being crucial to the development of the U.S. technopolis: the research university; large technology companies; small technology companies; national, state, and local government; and support groups (see Figure 2.1). They also stressed the importance of influencers or visionaries who network across these segments to foster cooperation among what are traditionally competitive public and private sectors. This chapter begins to document the generalizability of technopolis theory to the European setting.

The creation of CSATA (Centro Studi Applicazioni in Tecnologie Avanzate) was the first step in forming Tecnopolis Novus Ortus. CSATA was initially located on the campus of Bari University and was closely affiliated with the Computer Science and Physics Departments. In 1984 CSATA moved into its present facility at Tecnopolis Novus Ortus.[2] This emerging technopolis is modeled after the Japanese technopolis concept, which stresses locating high-tech research facilities out of congested urban areas (Tatsuno 1986).

In 1988 CSATA changed its name to Tecnopolis CSAT Novus Ortus (from now on TCNO) in order to stress its link to the technopolis concept. TCNO is situated near Valenzano, a small town about 15 kilometers from Bari in the Apulia region—which has traditionally been an important meeting place for the people and cultures of the Mediterranean basin. The Apulia region has about 4 million inhabitants, two universities (Bari, and Lecce) with a combined student population of 63,000, and 16 public institutions and research centers. In 1951 there were 753,000 Apulians involved in agricultural and 494,000 involved in the industrial and services sectors. By 1985 the agricultural sector had decreased to 319,000 persons, and the industrial and services sectors had grown to 985,000 persons.

TCNO offers business people throughout the Mediterranean the opportunity to take advantage of the technopolis strategy of collaboration between advanced and developing countries. For example, Tecnopolis Novus Ortus is the seat of the Community of Mediterranean Universities (CUM) which brings together more than 80 universities in projects aimed at scientific and education cooperation. And it has been instrumental in setting up links with large concerns operating in high technology such as IBM Italia, Olivetti, Italtel, Sip, Fiat, CCC (Computer Curriculum Corporation), Datanet, Dataconsyst, RLG (Research Library Group), Incorporated, and Sesa Italia.

A major goal of Tecnopolis Novus Ortus—Italy's first science park—is to spur high-tech economic development in the southern part of the country. Traditionally, Italy's economic growth has been in the north, in such areas as Milan and Turin. Political, industrial, and academic leaders in general have long viewed the south as coming in a poor second with regard to education and industrial strength. Most of the large Italian companies have their headquarters in the north and the most prestigious Italian university is in Milan.

The developmental gap between the southern and the northern industrial areas is demonstrated in the per capital income, which averages about $100 for the north and $68 for the south. However, the northern industrial regions are experiencing problems that inhibit continued industrial development. First, there is limited room to expand industrial facilities. Second, the demand for skilled workers exceeds the available supply. For example, the demand for electrical engineers is double the university output.

Tecnopolis Novus Ortus (TCNO) was designed in the 1970s with the goal of a campuslike facility. The strategy was a very pragmatic, step-by-step approach. There was not a lot of formal planning. The idea was to create, on a small scale, all the functions and amenities thought to be important to a larger technopolis. These facilities include a state-of-the-art conference room, a restaurant, recreational facilities, office and lab space, an international well funded library, data banks, and educational facilities. The initial facility was 11,000 square meters on four hectares. There are currently six attached buildings. As of 1988, TCNO—the consortium running the park—employed 200 professionals and staff. By 1990 this is expected to grow to 250 employees. In a sense, Tecnopolis Novus Ortus is a two-tiered organization: one of professionals and one of clerical support.

In many ways, Tecnopolis Novus Ortus exhibits an unusual atmosphere for an organization based in southern Italy. It seems to visitors more like a Silicon Valley high-tech firm than an indigenous Italian organization. This technopolis is an oasis in the dry Bari landscape. The facility's amenities include fountains and landscaped gardens as well as tennis and handball courts. The one large on-site restaurant, for all employees, provides an ample selection of quality food. For after-meal relaxation and informal get-togethers, there is an espresso bar with pastry, ice cream, and other snacks. Such amenities are rare—if they exist at all—among Italian firms. Perhaps the preferred culture of high-tech firms or technopoleis transcends national boundaries.

Employee ties to Tecnopolis Novus Ortus are strong. There are many instances of employees who moved north and who soon realized that they wanted to return to Tecnopolis—in part because it is an exciting place to work, with its unique mission and culture; in part because of Bari's climate and life-style; and in part because of strong ties to family in the Bari region.

Tecnopolis Links to the University of Bari

The links of Tecnopolis Novus Ortus to the University of Bari center on the students. Graduate and postdoctoral work on specific projects at Tecnopolis Novus Ortus in cooperation with the permanent staff of TCNO and company representatives from established Italian firms as well as local small and medium-size firms. Seventy-three recent postdocs currently work on research projects at TCNO. The affiliated companies support the students with hands-on training in company-sponsored projects. Examples of these programs are (Figure 3.1):

• TCNO–Centro Ricerche Fiat (CRF). The collaboration between TCNO and CRF—begun in June 1986—concerns the field of advanced robotics. The project's objective is the design of a "system of vision based on knowledge."
• TCNO–Olivetti. The collaboration between TCNO and the DOR (Direzione Olivetti Ricerche) laboratory—started in May 1986—has the objective of processing multimedia documents for office automation.
• TCNO–Italsiel. The collaboration between TCNO and Italsiel began in June 1986. Italsiel is the biggest Italian software house and one of the biggest in Europe. This collaboration regards the field of education. It concentrates on training executives, making computer-aided instruction (CAI) products for banks, and the development of a methodology related to the production of CAI software.
• TCNO–Laben. Laben works on aerospace robotics and is part of the multinational company ISC. Laben opened an office in TCNO in May 1986. The collaboration between Laben and TCNO aims at increasing the market and the availability of artificial intelligence technologies and of advanced automation for future European space missions.
• TCNO–Fiar. The collaboration between TCNO and Fiar started in October 1987. The objective is to create a center for research, development, and training in the field of robotics and artificial intelligence for the application of these technologies to space and submarine systems.
• TCNO–Telettra. An advanced group from Telettra was set up in 1987 at Tecnopolis Novus Ortus. Telettra is a telecommunications products manufacturer and is part of the Fiat group. The TCNO-Telettra collaboration will result in the creation of a software laboratory to be used in telecommunications.

Students enjoy taking courses that use TCNO's latest technology (e.g., robotics and computer support), which is generally not available at their universities. Interestingly, some of these students are from more technologically advanced northern Italy. The programs that most attract these students to Tecnopolis Novus Ortus are:

• SPEGEA (Business Management and Training School). A nonprofit consortium created by the Industrial Association of Bari with the collaboration of the Bari Chamber of Commerce, National Industry, Crafts,

Figure 3.1   Research Programs at Tecnopolis Novus Ortus (TCNO) That Include the Participation of Major Italian Firms

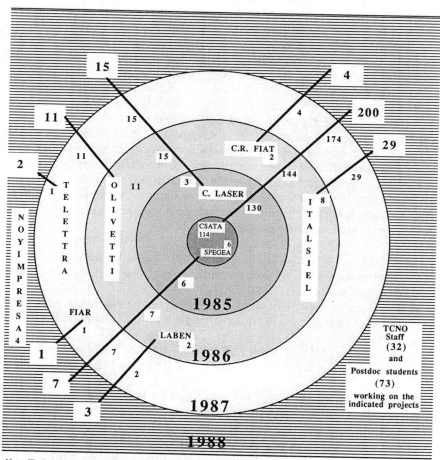

Note: The location of the company names indicates the start of the collaborative project. The numbers located next to the projects indicate the number of company employees working on the project in residence at TCNO, by year.

Source: The Authors.

and Agriculture, as well as a number of local firms. SPEGEA promotes the advanced training of management personnel for private and public companies and service organizations. The technical-scientific committee that governs SPEGEA is a multidisciplinary group of academics from a range of Italian universities as well as representatives from industry.

• SASIAM (School for Advanced Studies in Industrial and Applied Mathematics) is an international center for training and research in industrial mathematics. SASIAM is structured to be a bridge between university and industry in that the main objective of the workshop is to bring together researchers from both the academic and industrial areas—and thus stimulate the transfer of the mathematical experience to the solving of industrial problems. The researchers and students exchange ideas and discuss problems encountered in the formulation, analysis, and computer simulation of mathematical models of industrial processes.

• ASMIT (Advanced School for Mastering Information Technology) is an international school for training executives in the integration of financial and administrative skills with technical and scientific knowledge. ASMIT has three major objectives:

1. A master's program combining the application of new information technologies with innovative processes for business formation and development;
2. An Executive Development Center for senior managers who need to make strategic decisions involving advanced information technologies; and
3. A Business and Technology Support Center for young entrepreneurs to aid in the creation of new enterprises in information technologies.

Tecnopolis Novus Ortus offers two kinds of fellowships: (1) ten one-year postdoctoral positions for focused research; and (2) ten three-year postdoctoral positions that include one year of study abroad in the United States, Canada, or Europe. The preferred subject of study is microelectronics. These postdoctoral students receive credit for their work and are often hired later by the companies associated with TCNO.

Tecnopolis Bari's links with university professors are more problematic. An obstacle to cooperation with the University of Bari concerns the conflicting perceptions of who is most qualified to head research and education programs at Tecnopolis Novus Ortus. It has always been the policy of the general director of TCNO to insist on hiring the highest quality professionals possible from any university or region in Italy. On the other hand, some senior professors at Bari University believe that—

based on their seniority at the university and their institutionalized importance, or their perception of the importance of their own academic work—they should have the prestige of heading the TCNO programs.

Such a conflict of perceptions puts TCNO in a very delicate position in its dealings with the university. Nevertheless, the University of Bari has two nationally recognized schools, and there are cases where senior Bari professors have been chosen by TCNO to head a major research effort.

Links with State and Local Government

The initial 1984 funding for Tecnopolis Novus Ortus came from Italy's national government. A key motive was to encourage economic development in Southern Italy. As of 1988, Tecnopolis Novus Ortus had an operating budget of 18 billion Italian lira, which included such general expenses as salaries for 200, facility and equipment maintenance, and software, hardware, and library costs. The role of the national government in the initial development of Tecnopolis Novus Ortus has been crucial. Also, a number of TCNO activities have been commissioned by local government administrators as a basis for further collaboration.

Tecnopolis Novus Ortus Links with Italian Industry

Tecnopolis Novus Ortus has both local and national ties with Italian industry. The technopolis is in the business of selling its expertise to large Italian industry as well as small start-up companies. At the level of larger more established industry, the operating concept of TCNO is to form small work units that are involved in R&D activities of practical interest to the company. Research teams are usually made up of five or six professionals from the company, TCNO and the University of Bari (Figure 3.1). For example, a large company in Milan will send a senior researcher to Tecnopolis Novus Ortus who will act as the team leader of a work group made up of two or three TCNO professionals and two to four recently graduated Ph.D. students. The company defines the project objectives in discussions with a representative from TCNO. It is important that the results of the research be of use to the company. A typical program lasts three years and is subject to yearly monitoring. Often at the end of the project the company will recruit the Ph.D. graduates who worked on it.

Seed/venture capital is almost nonexistent in Italy, so typically new ventures are funded by the entrepreneur's own cash reserves and the bonus/savings he or she receives on leaving the "mother" company. According to the General Director of TCNO, Italy is bound by an antiquated financial point of view: Banks want exceptional guarantees for their loans. The country needs financial consultation with regard to restructuring its system to encourage more entrepreneurial activity.

A goal of TCNO is to get major Italian firms to set up their own entrepreneurial activities in Bari to assist start-up or spin-off companies. In this regard, there is an interesting new form of entrepreneurial behavior being exhibited by Italian senior managers who are not satisfied with the operating policy or strategy of their company. In short, the senior manager proposes to start a new company. Top management says no, and the manager takes the idea and goes it alone. There have been three recent cases at Tecnopolis Novus Ortus of such spin-offs.

Netsiel, a software firm and a spin-off from Italsiel, was one of TCNO's first spin-off companies. Another spin-off company that located at Tecnopolis Novus Ortus in 1987 is Telettra. Partly as a result of these spin-off activities, Bechtel Corporation, U.S.A., was hired in 1988 as a consultant to help plan for the expansion of the Bari technopolis.

One of the problems currently facing TCNO is managing the real estate associated with the growing technopolis. Since Tecnopolis Novus Ortus does not own the adjoining land, the existing landowners have benefited tremendously from TCNO's development, in that their land has increased in value. In response to this rapid growth, it has been recommended that TCNO create a real estate company to oversee the development and maintenance of the emerging technopolis.

Spin-offs of the Technopolis Concept throughout Southern Italy

Tecnopolis Novus Ortus has inspired five spin-offs of the technopolis concept in Southern Italy: CORISA, CRAI, CRES, CRIAI, and CSATI. In 1988 all five spin-offs were about where CSATA had been in the 1970s, and they continued to see CSATA (now TCNO) as a role model. The vision of each of these spin-off programs originated within the region's local universities. While there was little tangible support at each university, usually one senior professor was key in developing and championing the concept of a local technopolis. And in every case there followed the recruitment and training of young people and the creation of an organization on the model of Tecnopolis Novus Ortus.

In November 1987 all six (including TCNO) of these emerging technopoleis created a second-level consortium called IATIN. The president of IATIN is from TCNO, and the vice-president is from CRES. Members of IATIN communicate by a computer network—the only such network in Southern Italy. Indeed, there is nothing like it in Northern Italy. The goal of the participants is to form a common understanding among themselves of the technopolis concept and to present this common vision to Italy's industrial leaders and to the EEC (European Economic Community). There are R&D people in all six technopolis sites working on some 500 projects in such areas as artificial intelligence, communications, remote sensing, robotics, software, and microelectronics.

URT (United for Technology Transfer) is concerned with technological diffusion at the regional level—primarily for small (less than 100 employees) and medium-size (less than 500 employees) companies. URT, funded by the Agency for the Development of Southern Italy, was started in 1987 and is currently running in three emerging Southern Italian technopoleis—CRAI, CRES, and CRIAI—as well as TCNO.

In promoting technology transfer at the regional level, URT generally follows the procedure depicted in Figure 3.2 and described below.

1. Hold a one-day seminar on an important, current technology-related topic to attract people from small to medium-size firms and local government officials. Encourage general discussion. URT is based on the idea that it is important to educate local government officials, as well as the private sector, to the dynamics of technology transfer. It has been found from experience that local government agencies can do much to either block or foster technology transfer and the economic development of a region.

2. Identify two to four key people at the seminar who exhibit a positive response to the presentation of the innovative technology.

3. URT representatives then propose an experiment with these key people that consists of moving personnel and technology from TCNO into the company. For example, a TCNO experimental engineer would work on location with a company team on a company's CAD problems for two or three months. The company is expected to commit itself and really participate and work with the TCNO representative. TCNO identifies the benefits and issues of the technology-transfer attempt in a report to the company. At this point in the process, TCNO's support stops. TCNO has no vested

Figure 3.2   The URT Model to Promote Technology Transfer to Small and Medium-size Firms

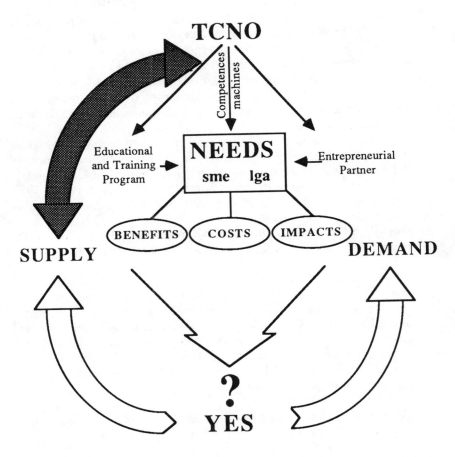

sme = small and medium sized enterprises

lga = local government agencies

Source: The Authors

interest in pushing a particular product. The goal is to help the company. The company makes its own choice concerning hardware and software purchases and may even ask another consulting company for continued support concerning technology purchasing, implementation, and operation.

A project called STAR (Advanced Development of a Region through Advanced Communication Services) has just received support from the EEC to replicate the URT concept throughout the less economically favored regions of Europe: Southern Italy, Greece, Portugal, Spain, Ireland, and parts of the U.K. and France (Figure 3.3).

The Future

At the end of 1988, the development of Tecnopolis Novus Ortus received an important boost following approval by the Italian central government to finance the next three years of TCNO activities. The funds

Figure 3.3   **STAR, the URT-based Model for Promoting Technology Transfer to Less Developed Regions within the European Economic Community**

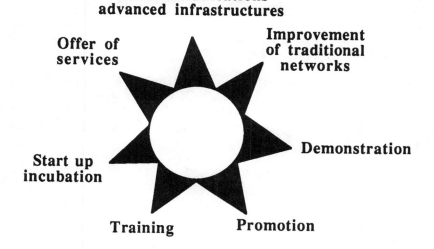

Telecommunications
advanced infrastructures

Offer of services

Improvement of traditional networks

Start up incubation

Demonstration

Training   Promotion

allocated to this venture amount to 135 billion Italian lira. This includes funding for new buildings and laboratories as well as support for existing and new projects.

Bechtel Civil Engineering, U.S.A., has been commissioned to draw up a feasibility plan for TCNO during the next three years. This growth planning coupled with numerous other projects from public administration and from the European Community poses new challenges for Tecnopolis Novus Ortus. Mainly, TCNO must manage its own institutional growth and yet not lose sight of its primary goal: local socioeconomic development. The experience it has accumulated and the links set up with the University of Bari, small and large technology companies, and national and local administration attest to the viability and the generalizability of the technopolis concept to the European setting.

Notes

1. A technopolis is a property-based initiative that has links with universities and research centers and aims at promoting local socioeconomic development through technology transfer.

2. CSATA (Centro Studi Applicazioni in Tecnologie Avanzate) is the name of the consortium that originated the idea of a technopolis in Southern Italy. Tecnopolis Novus Ortus is the name of the technopolis created by CSATA. Tecnopolis CSATA Novus Ortus (TCNO) has been the name of CSATA since 1988, and it identifies the consortium in charge of running the first Italian technopolis, which is located in Bari, Italy.

References

Raymond W. Smilor, G. Kozmetsky, and D. Gibson, eds., *Creating the Technopolis: Linking Technology Commercialization and Economic Development* (Cambridge, Mass.: Ballinger Publishing, 1988).

Sheridan Tatsuno, *The Technopolis Strategy* (Reading, Mass.: Addison-Wesley Publishing, 1986).

Part II

University Faculty as Entrepreneurs

4

Exploring Perceived Threats in Faculty Commercialization of Research

David N. Allen and Frederick Norling

U.S. industry is restructuring along many fronts. It is moving to smaller scale, flexible processes that quickly respond to customers' needs. In goods and services production today, value is primarily added by information, in the form of knowledge. Although colleges and universities are at the leading edge of knowledge creation, seldom have they played a role in translating that knowledge into later stages of product development or actual product commercialization.

The faculty is the embodiment of the university. Thus, faculty members are the key players in the knowledge creation process. This chapter departs from the role of faculty in knowledge creation to examine its role in knowledge transformation. Interest focuses on the transformation of research to commercial value. Three generic transformation functions are analyzed: (1) client-based research; (2) consulting; and (3) business start-up. These activities are examined for their perceived threats to the traditional culture of the United States.

The intellectual world operates independent of the commercial world. Intellectual or research products prosper in a nonintervention culture that permits freely interacting researchers to seek consensus and pursue new inquiry (Burke 1986, 133–54; Roy and Shapley 1985). Often, multiple solutions are posed—with the uncertainties emphasized. Peer review and the respect of colleagues are the criteria for success. Conversely, commercial products rely on a culture where proprietary interests are protected and become competitive advantages. Single solutions that maximize profit are posed—with the uncertainties for users resolved. Market acceptance and profitability are the criteria for success.

Seldom do the intellectual world and the commercial world influence one another, except in an indirect manner. Universities change at glacial speed, and perhaps this is why they are some of the most enduring institutions in Western society. Even they, however, are feeling the pressures for change (Dimancescu and Botkin 1986; Lynton and Elman, 1987, 16–29). The external support environment is different now—what with the decreasing young-adult market, fiscal austerity, and the quiet but growing demand by state governments for economic returns on public higher education expenditures. These forces have prompted academic administrators to create and revise programs that establish supportive linkages between industry and academe. It is thought that in the long run the new environment will help satisfy needs in research, instruction, service, and funding areas.

Universities and Economic Development

The response of universities to today's internal and external pressures is in the direction of wealth creation (economic development), although the traditional emphasis on knowledge creation will continue to play a primary role. In a recent survey of nearly 300 public universities, 97 percent had plans to increase their economic development activities (SRI International 1986). The approaches toward wealth creation are varied, however.

The first economic development approach coincides with the traditional comprehensive education objectives of developing human resources. Studies have shown that industry values higher education for its output of trained, motivated, low-salary workers—graduates (Allen and Levine 1986, 119–50). New programs in artificial intelligence, computer engineering, entrepreneurship, and so forth are currently creating a supply of cutting-edge employees. Expansion of other comprehensive educational programs such as continuing education and professional management training can provide the community with a means to update skills and inculcate lifelong learning.

The second approach—research—also fits within the traditional bounds of higher educational pursuits. Industry-oriented research has greater potential for commercial application than basic research that is done more for the sake of knowledge. The new organizational arrangements such as research consortia, centers of excellence, and cooperative research centers tend to move faculty research from a knowledge-based interest to a proprietary-based interest focused on product and process development concerns (Ryans and Shanklin 1988; Colton 1985; Norling and Allen 1986).

Technical and managerial assistance specific to a client or a well-defined community of interest is the third type of economic development approach. Universities are giving increased attention to assistance arrangements where the expertise of individual academics can be directly brought to bear on particular problems. Three common technical-assistance approaches include agricultural extension, industrial extension, and faculty consulting. On the business management side, universities in cooperation with states and the federal government have created extensive small-business development centers, small-business institutes, and training and consultative organizations (McMullan, Long, and Graham 1986; Steinnes 1987; Christman, Hoy, and Robinson 1987).

A fourth area of university involvement in economic development occurs at the site or facility level. Companies have connected themselves with universities for access to sophisticated laboratories, computers, and other capital-intensive facilities. The past decade has witnessed a rapid growth in technology and research parks affiliated with universities. More recently, nearly 50 institutions of higher education—individually or in partnership with public and private organizations—have developed business incubators (National Business Incubation Association 1988).

Fifth and last, many university administrations have sought to develop a positive environment for commercial pursuit on campus. Activities in support of this mission include all of the four previously mentioned approaches plus attention to creating flexible, supportive faculty reward structures. At the heart of the controversy surrounding all these approaches (some more than others) is conflict of interest and conflict of commitment, as related to faculty proprietary endeavors. Few institutions have actually stepped out and drastically changed the criteria for promotion and tenure to include commercial pursuit, but commercial activities are now seen more positively than in the past. Some universities openly support faculty members who desire to develop their research commercially through independent companies (Brock 1985; Smith, Drobenstott, and Gibson 1987).

Faculty Commercial Activities

Three new kinds of faculty activity characterize the shift of universities in the direction of economic development. We may place these activities on a three-by-three matrix according to their degree of proprietary interest and general-versus-specific knowledge application (Table 4.1). *Client-based research* is the contractual obligation of faculty to pursue inquiry that is expected to have commercial value. This kind of activity has midrange proprietary interest and is usually predevelopmental. Developmental continuation of this research typically occurs under the guidance of the client or whoever will be receiving rights to the intellectual property. *Faculty consulting* is more problem solving than developmental and involves a relatively high proprietary interest. Faculty consulting is important because it gives faculty members the opportunity to work with industry counterparts (Eveland 1985). It is often a precursor to future arrangements between researchers and their corporate partners (Fusfeld 1983; Lynn 1984). *Faculty start-up* is the epitome of high proprietary interest and specific knowledge application. By starting a business related to his or her research endeavor, the faculty person: (1) retains control of the intellectual property; and (2) creates the possibility for high upside personal economic reward. If that faculty person chooses to remain at the university, a host of conflicts may ensue; but at the same

Table 4.1   Faculty Commercialization Activities

	Knowledge Application			
Proprietary Expectations	Basic Research	Applied Research	Problem Solving	Production
Low	National Institute & Foundation Supported Research			
Medium		Client Based Research		
High			Faculty Consulting	Start-up

Source: The Authors

time—by staying in academe—he or she is protected against the risks associated with business failure. Staying or leaving may also have implications for the long-term success of the company (Doutriaux 1987).

All three of these new activities challenge the university's traditional culture of conservatism and resistance to change. It is common for "faculty members to take the position that universities should be detached and objective observers and critiques of the world around them, not active participants in it" (SRI International 1987). The new activities gain prominence when critics claim that they distort institutional priorities, impair teaching or other academic duties, weaken objectivity, constrain publication and information dissemination, or entice faculty to leave academe (Government–University–Industry Research Roundtable 1986, 22–23; Allen and Bird 1987). Of these five claims, three can be examined directly by using data collected from 396 faculty members at 37 universities in Pennsylvania. The research hypotheses (stated in null form) are:

Client-based research, consulting, and faculty start-up activities do not affect
1. institutional priorities.
2. teaching or other important activities.
3. a desire to leave the university.

Methodology and Sample

All four-year higher education institutions in Pennsylvania were used in the design of this sample. Forty of the 135 in the population of higher education institutions were randomly selected. Random selection ensured that large and small, research and nonresearch, and state and private schools had an equally probable chance of being selected. Three schools chose not to participate. A sample of faculty proportional to the size of the institution was drawn from institutional directories. A total of 912 faculty members from schools, programs, departments, or colleges of medicine, business, science, and technology comprised the prospective pool of respondents. These people were sent a four-page survey during the spring of 1986. A follow-up survey was sent a few weeks later. All told, 440 individuals responded. Forty-two surveys were returned blank or unusable. The number of cases used for analysis was 398, which amounts to a 43.6-percent response rate (see Table 4.2).

The sample is comprised almost exclusively of advanced-degree hold-

Table 4.2   1986 Survey of Industrial Contact Activities among Pennsylvania Faculty: Population and Sample Characteristics

Count (percent)	Total	Science	Technology	Business	Medical
Population	4050	1827	827	748	648
Sample Population	912 (22.5)	398 (21.8)	188 (22.7)	176 (23.5)	150 (23.1)
Final	398 (43.6)	185 (46.5)	99 (52.7)	60 (34.1)	54 (36.0)

Source: The Authors

ers: 79.9 percent, Ph.D.s, 9.5 percent, Masters, and 5.1 percent, M.D.s. School-based affiliations are greatest for science, 46.5 percent; and technology (including engineering), 24.9 percent; and least for business administration, 15.1 percent; and medicine, 13.6 percent. The mean number of years employed at the institution is 13.4, with a mean of 16.9 years since award of the terminal degree. Nearly two-thirds (64.4 percent) are tenured. This sample of faculty respondents is similar in many ways to groups studied elsewhere.[1]

Classification of Faculty

University faculty are not monolithic; job requirements, areas of expertise, and interests vary considerably. As part of the Pennsylvania survey, faculty members were asked to indicate whether they pursued any of eight kinds of industrial contact activities emanating from their research. Respondents were asked what kind of private-sector contacts they had engaged in over a six-year period (1983–88), as well as whether they desire to have such a contact. Of the eight activities, three are of particular interest for this analysis: client-based research, private consulting, and business start-up. To create a mutually exclusive, exhaustive classification variable that would include the combination of activities, a seven-value measure was created.

The first and largest group (110) is composed of respondents who did not conduct research or have contact with private-sector clients.[2] The second group (62) is comprised of faculty members who did have private-sector contacts, but not as part of client-based research, personal consulting, or a business start-up. These individuals had engaged in industrial contact activity either as unpaid advisors, or during sabbatical leave, or while supervising graduate students, performing institutional consultancies, or conducting non–client funded research.

The third group of faculty (27) includes those who only do client-funded research—that is, no consulting or start-up activities, or desire to do so. The fourth group (48) is comprised of those who do just private consulting—no client research or start-up. The fifth group (13) is made up of those who have engaged or desire to engage in start-up activity and either private consulting or client research. Respondents in the sixth group (85) do client-based research and private consulting. In the seventh and last group, 51 faculty respondents do all three: client research, consulting, and start-up.

The frequency distribution of this last variable provides an interesting perspective on how start-up activity fits into the wider realm of faculty activity. No faculty member reported that he or she was involved in a start-up independent of client research or private consulting. Quite the contrary is true; start-up activity is imbedded within other client-based endeavors. Where start-up activity or interest occurs, it is with either research and consulting (51), or research or consulting (13). This is one indication that start-up activity is woven into the fabric of widely accepted activities (such as client research and consulting). The finding also suggests that start-up activity occurs within the context of industrial contact. Last, it indicates that start-up activity is the province of only a few. The two groups that include start-up activity constitute only 16.2 percent of the entire sample. If one subsets just those who have actually engaged in start-up activity based on their research, the percentage of the entire sample falls to 4.4 percent.

Bivariate relationships between the faculty classification variable and certain other commercialization variables shed some light on the unique nature of faculty start-up interests. Respondents were asked if they hold or plan to hold a patent. Slightly more than one-third of the faculty members with a start-up interest said that they have or expect to have a patent. Conversely, 23 percent of the research-only faculty and 16.7 percent of the consulting-only faculty have or expect to have a patent. A series of questions asked whether or not the respondents had been approached by private-sector parties to: (1) join in a partnership; (2) license their technology; (3) manufacture their product; (4) finance their new business; or (5) hire them. The percentage of positive responses to every one of these five items was more than double or greater for faculty with a start-up interest compared to the other four applicable groups of respondents.

Challenges to Teaching and Academic Activities

Outsiders are often puzzled about how university faculty spend their time. Almost all faculty members have teaching responsibilities; but at research institutions, research productivity is more highly valued than instructional productivity. Critics of university enterprise contend that attention given to commercialization activities will detract from the time spent on instruction, research, and service activities (Government–University–Industry Research Roundtable 1986, 22–23). They say that this diffusion of time and interests will have a negative effect, eventually undermining the foundations of university support.

To examine this contention, we asked our survey respondents to indicate the percentage of time they devote to each of five activities (Table 4.3). Almost all of our faculty members participated in resident instruction (94.9 percent). They spend nearly half of their time (47.2 percent) in this activity. Slightly greater than four-fifths (81.5 percent) are involved in university service or service to a particular community of interest (government, professional society, and so forth). Among those engaged in service, about 17 percent of their time is spent in this activity. About three-quarters (76.5 percent) do scholarly activity, spending about 22 percent of their time in this mode. Greater than half the sample (54.4 percent) conduct sponsored research. This activity takes about 35 percent of their time. Approximately 14 percent of the entire faculty engage in continuing education activities and spend an average of 30 percent of their time thusly.

Except for a few instances, the time allocations seem to vary little by classification of faculty type (Table 4.3). The single greatest difference is between the nonresearch faculty and all others. Nonresearch faculty members spend about two-thirds of their time on instruction, while everyone else spends from about 40 percent to about 50 percent of his or her time on this activity. Moreover, for these four other applicable groups, the average and percentage differences between groups are negligible and not statistically significant at the .01 level.

The percentages and averages contained in Table 4.3 strongly indicate that faculty members who engage in industry-related activities such as contract research and consulting are as actively involved in important

Table 4.3 1998 Survey of Industrial Contact Activities among Pennsylvania Faculty: Time Allocation of Faculty Classification

Count (% of row) Mean Number of Hours		Resident Instruction	Continuing Education	Sponsored Research	Scholarly Activities	University Service
Non Research	110	106 (96.4) 62.0	34 (30.9) 18.5	15 (13.6) 38.3	77 (70.0) 18.7	83 (75.5) 18.7
Neither Research Consulting nor Start-up	62	58 (93.5) 42.5	12 (19.4) 6.5	35 (56.5) 40.3	49 (79.0) 26.6	51 (82.3) 17.0
Research Alone	27	23 (85.2) 43.3	8 (29.6) 24.6	20 (74.1) 32.6	17 (63.0) 19.6	17 (63.0) 18.8
Consulting Alone	48	44 (91.7) 43.2	13 (27.1) 13.2	25 (52.1) 31.6	40 (83.3) 28.3	40 (83.3) 17.6
Research and Consulting	85	83 (97.6) 51.2	28 (32.9) 29.0	75 (88.2) 28.8	70 (82.3) 18.5	78 (91.8) 18.3
Start-up and 1 or the other	13	12 (92.3) 51.2	5 (38.5) 29.0	4 (30.8) 28.8	11 (84.6) 18.5	12 (92.3) 18.3
Research, Consulting and Start-up together	51	47 (92.2) 40.3	20 (39.2) 11.9	39 (76.5) 36.8	37 (72.5) 21.5	40 (78.4) 15.7
Total	396	373 (94.4) 47.2	120 (30.6) 13.8	213 (54.4) 35.2	301 (76.5) 22.2	321 (81.5) 16.9

Source: The Authors

aspects of university life as their non–research oriented colleagues. Faculty members who do not conduct research spend more time teaching. However, all other faculty members are as likely as the nonresearch faculty to be involved in some resident instruction. Nonresearch faculty members are not appreciably more involved in other noninstructional activities nor, when they do become involved in other activities, do they allocate a greater proportion of their time to those activities. The group of "supercommercial" faculty—that is, those who have client-based research, consulting, and start-up interests—are good academic citizens: They are as involved in university activities as other faculty, and they tend to allocate a comparable amount of time to these activities.

Challenges to Institutional Priorities

A college or university is a community of scholars assembled for the pursuit and application of knowledge. The personal and professional goals of the members constitute the priorities for the institution. Higher education in the United States has a strong tradition of personal autonomy and discretion among faculty (Lynton and Elman 1987). One purpose of academic tenure is to isolate faculty members from pressures that may impair their academic integrity and credibility. Some fear that the increased attention on commercial endeavor will change our most basic value assumptions about the primacy of research and instruction.

This perceived threat to traditional values can be examined via faculty responses concerning the relevancy of 11 key personal and professional goals. For each goal, respondents were asked to judge relevancy on a five-point scale with 1 anchored at irrelevant; 3, neutral; and 5, relevant. The mean and relative ranks for each goal according to the faculty classification variable are presented in Table 4.4. Generally speaking, the faculty groups judge relevancy in a similar manner. The least similar group is composed of those individuals who have start-up interests and are engaged in either consulting or client research. It should be noted that this is the smallest group ($N = 13$, or 4.5 percent of the sample), and small groups are prone to sample fluctuations.

For the five remaining faculty groups, the most important goal is "publish in scholarly sources." Ranks two and three are consistent across the five faculty groups; both goals relate to the increase of knowledge: "pure scientific" and "applied/problem solving." Goals ranked fourth and fifth—"improve education for graduate students" and

Classification

Scale Anchors: 1=irrelevant, 3=neutral, 5=relevant

Source: The authors.

Mean, Rank	Publish	Pure Knowledge	Applied Knowledge	Collaboration with Academics	Graduate Student Education	Increase Income	Social Benefit	Develop Methodology	Develop Class Materials	Develop Product/Process	Establish own Firm
Neither Research Consulting nor Start-up N=62	4.5 / 1	4.5 / 1	4 / 3	4 / 3	3.6 / 5	3.2 / 9	3.3 / 7	3.3 / 7	3.4 / 6	2 / 10	1.4 / 11
Research Alone N=27	4.5 / 1	4.3 / 2	4.3 / 2	4.1 / 4	4.1 / 4	3.4 / 7	3.6 / 6	3.2 / 8	3.2 / 8	2.9 / 10	1.3 / 11
Consulting Alone N=48	4.3 / 1	4.1 / 2	4 / 3	3.8 / 4	3.5 / 4	3.4 / 7	3.2 / 6	2.8 / 7	2.8 / 9	2.6 / 9	1.4 / 11
Start-Up and Consulting N=13	4.5 / 4	4.3 / 2	4.2 / 1	3.9 / 9	3.9 / 11	3.3 / 2	3.4 / 8	3.3 / 4	2.9 / 6	2.9 / 7	1.3 / 10
Research, Consulting and Start-up Together N=51	4.6 / 1	4.4 / 3	4.5 / 2	4 / 5	4.2 / 4	3.7 / 7	3.4 / 9	3.6 / 8	3.1 / 10	3.9 / 6	2.6 / 11
Overall Mean	4.4 / 1	4.3 / 2	4.2 / 3	3.9 / 4	3.7 / 5	3.4 / 6	3.3 / 7	3.3 / 7	3.1 / 9	2.6 / 10	1.6 / 11
N for Overall Mean	289	292	290	290	287	288	286	288	289	283	287

"collaborate with other academicians"—are virtually the same across the five faculty groups.

In a manner paralleling the five most relevant goals—although not as consistent—the five least relevant goals are similar across the five faculty groups. "Establish own firm" is the least relevant goal (and second least relevant for the group involved in start-up and one other activity). Also included in the five least relevant are "develop methodology/instrumentation," "develop classroom materials," "increase income/employment opportunity," and "apply knowledge for social benefit." One difference among the least relevant goals concerns the group that does not become involved in consulting, client research, or start-up; this group places greater importance on the development of classroom materials. (Recall that they also spend a greater percent of their time in resident instruction.)

Another difference concerns the two groups involved in start-up activities. Generally, these two groups are slightly more likely to respond to pecuniary concerns such as "developing a product or process" and "increase income/employment opportunity." However, both start-up groups place low priority (relative to the other ten goals) on "establish own firm."

This analysis is consistent with the previously stated patterns that show the supercommercial faculty to be tied to the norms of academic life. Although they may not be clones of the other groups involved in consulting and client research activity, the supercommercial faculty exhibit a goal structure that should not threaten the traditional culture of the university.

Challenges to Faculty Retention

One of the most damning and direct criticisms of increased industrial client contact is that it creates a temptation for faculty to leave academe. Universities commit substantial resources to attracting and retaining commercially relevant faculty. In recent years, more positions in engineering and business were available than were filled (Bowen and Shuster 1986). Furthermore, if a university loses a faculty member, it also stands to lose graduate students, research aides, and other support staff and facilities that the faculty person has attracted to the university.

Our faculty respondents who were able to do so were asked whether the commercial potential of their research could possibly motivate them to leave academe (Table 4.5). Of the respondents involved with industrial contacts, 28.3 percent said they could possibly leave university life. The

Table 4.5 1986 Survey of Industrial Contact Activities among Pennsylvania Faculty: Possibility of Leaving Academe, by Faculty Classification

Frequency (row percent)	Possible Leave	Not Possible Leave	Total
Neither of three	5 (10.4)	43 (89.6)	48 (17.8)
Research Alone	6 (23.1)	20 (76.9)	26 (9.7)
Consult Alone	8 (17.0)	39 (83.0)	47 (17.5)
Research and Consulting	27 (31.8)	58 (68.2)	85 (31.6)
Start-up and one or the other	6 (50.0)	6 (50.0)	12 (4.4)
Research, Consulting and Start-up Together	24 (47.1)	27 (52.9)	51 (19.0)
Total	76 (28.3)	193 (71.7)	269 (100)

Cramers V=.32
Chi Square prob=.001
17 missing cases

Source: The authors.

lowest chances of leaving are held by those faculty who do not have research, consulting, or start-up interests. Therefore—assuming an ordinal ranking of increased closeness to the industrial sector based on research, consulting, and start-up—the closer a faculty member moves to industrial clients, the greater the likelihood that he or she may leave academe. For example, nearly half of those involved in all three activities or start-up in conjunction with one other activity expressed the possibility of leaving. Conversely, of those involved in research or consulting alone, 23.1 percent and 17 percent—respectively—indicated the possibility of leaving.

The decision to leave the university is a difficult one. In many cases, it is irreversible: The university will not take back a former faculty member at the same rank and position, especially when the split has created ill feelings between the two parties. Respondents who said they could possibly leave the university were asked the reasons why they might. They were instructed to indicate any of four possible reasons that could motivate them to leave. Of the 76 who said they could possibly leave, 59 cited increase in income as a reason. Next most important (40 of 76) was the potential to expand their venture. The two least mentioned reasons were to gain greater ownership and to avoid legal entanglements (7 and 8, respectively). Seventeen gave other reasons that vary considerably, with no more than three persons claiming any one of them.

Summary

University faculties are often at the cutting edge of technology. Increasing attention is being given to the role of faculty in commercialization of their research. This chapter has examined faculty commercial interests and how these interests affect actual and intended behavior important to the traditional academic mission. A few general observations can be drawn from this analysis.

1. A low percent of faculty is interested in commercializing academic research through a start-up company.

2. Those interested in start-ups are also involved in client-based research or consulting. Usually faculty members interested in start-ups are involved in both these endeavors.

3. Faculty members with a start-up interest are much more likely to have been approached by private-sector parties to engage in collaborative commercial ventures.

4. Faculty members with commercial interests are involved in all aspects of university life. Even the supercommercial faculty group commits an average amount of time to instruction, scholarly activity, sponsored research, and service.

5. Little variation is seen in the ordering of personal and professional goals for faculty involved in different kinds and levels of commercial

activity. Some differences are seen in the saliency of various goals, but overall patterns are not evident.

6. Nearly 30 percent of our faculty respondents involved in commercial activities said they could leave academe. The greater the involvement with the commercial sector, the greater the likelihood of leaving the educational sector.

7. Increased income potential is a major inducement to leave university life.

Concerning the three research hypotheses posed earlier in the chapter, it seems safe to say that the first two have not been refuted. The weight of evidence clearly indicates that institutional priorities and important institutional activities are not threatened by commercial contacts and the commercialization of research. The third hypothesis is more problematic. In an era in which faculty loyalties are shared between the discipline, colleagues, and the institution (Government–University–Industry Research Roundtable 1986), it is perhaps not all that surprising to find nearly 30 percent of a faculty sample interested in leaving academe. This is especially so when pecuniary expectations are greater on the other side of the fence.

Conclusion

Its faculty is the linchpin in a university's movement toward expanding its traditional missions to include economic development. The faculty members conduct the research and create the knowledge that can be translated into commercial value. They are the embodiment of the university; for the academic climate to be receptive of industry's changing needs, faculty must be supportive of these new initiatives. It is the faculty that can produce the technically competent students sought by industry. Scholars also transfer their knowledge directly through consulting. In a few instances, faculty members have even been responsible for commercializing their own research through university spin-off corporations.

Acceptance of a change in university policy toward enhancing the entrepreneurial environment depends on the degree to which new initiatives enhance the fundamental missions of academe. As a general evaluative criterion for commercial initiatives, universities should examine how commercial value can be added to its traditional research, instruction, and service missions. To enhance the commercial and entrepreneurial environments, commercial endeavor and entrepreneurship must enhance the academic environment. This chapter indicates that universities have little to fear concerning commercial ventures taking away from their traditional missions. It will be up to the successful academic entrepreneurs themselves to show how commercial ventures can add back to the traditional missions, as well.

Most universities are just becoming involved in economic development; they are primarily in the planning—not implementation—stage (SRI International 1986). Because change at universities occurs through a consensus building process, it will be many years before most approaches are implemented and widespread results are evident. Anecdotal accounts of regional economic growth highlight the critical role a few universities have played in their own areas. It is becoming an article of conventional wisdom that when a university encourages its faculty and students to participate in entrepreneurial spin-offs, the local business climate improves.

Acknowledgments

The authors would like to acknowledge the assistance of Janet Hendrickson-Smith and Doug Smith for their tireless efforts in preparing the data. David Jordan and S. Sasmpathkumaran assisted with the data analysis. The Pennsylvania Ben Franklin Partnership Program administrators in Harrisburg were very helpful when we needed access to the universities. Ruby A. Sheperd did a skillful job in preparing the manuscript.

Notes

1. This sample is unlike the Rodenberger and McCray (1981) sample in that our study did not focus on founders who had commercialized their research as the core of their enterprise.
2. This group of faculty is used only for the time-allocation analysis. If respondents did not engage in research, they completed the time-allocation question but not the remainder of the survey.

References

D. Allen and V. Levine, *Nurturing Advanced Technology Enterprises* (New York: Praeger, 1986).

D. Allen and B. Bird, "Faculty Entrepreneurship in Research University Environments," in *Frontiers of Entrepreneurship Research 1987,* N. Churchill, J. Hornaday, B. Kirchhoff, O. J. Krasner, and K. Vesper, eds. (Wellesley, Mass.: Babson College, Center for Entrepreneurial Studies, 1987), pp. 617–30.

D. Birch, "Thriving on Adversity," *Inc.* (March 1988): 80–81.

H. Bowen and J. Shuster, *American Professors: A National Resource Imperiled* (New York: Oxford University Press, 1986).

D. Brock, "Faculty Business: Professors' Push for Profit Spurs Growing Concern," *Wall Street Journal,* August 27, 1985.

R. Burke, *Science, Technology, and Public Policy* (Washington, D.C.: Congressional Quarterly Press, 1986).

J. Christman, F. Hoy, and R. Robinson, "New Venture Development: The Costs and Benefits of Public Sector Assistance," *Journal of Business Venturing* 2, 4 (1987):315–28.

R. M. Colton, "Status Report on the NSF University/Industry Cooperative Research Centers," *Research Management* 28, 6 (1985):25–31.

D. Dimancescu and J. Botkin, *The New Alliance: America's R&D Consortia* (Cambridge, Mass.: Ballinger Publishing, 1986).

J. Doutriaux, "Growth Patterns of Academic Entrepreneurial Firms," *Journal of Business Venturing* 2, 4 (1987):285–97.

J. Eveland, *Communication Networks in University–Industry Cooperative Research Centers* (Washington, D.C.: NSF, 1985).

H. Fusfeld, "Overview of University–Industry Research Interactions," in *Partners in the Research Enterprise,* T. Langfitt, S. Hackney, A. Fishman, and A. Glowasky, eds. (Philadelphia: University of Pennsylvania Press, 1983), pp. 10–19.

Government–University–Industry Research Roundtable, *New Alliances and Partnerships in American Science and Engineering* (Washington, D.C.: National Academy Press, 1986).

J. Lynn, *The High-technology Connection: Academic/Industrial Cooperation for Economic Growth,* ASHE-ERIC Higher Education Research Reports (Washington, D.C.: National Institute of Education, 1984).

E. Lynton and S. Elman, *New Priorities for the University* (San Francisco: Jossey-Bass Publishers, 1987).

W. McMullan, W. Long, and J. Graham, "Assessing Economic Value Added by University Based Outreach Programs," *Journal of Business Venturing* 1, 2 (1986):225–40.

National Business Incubation Association, *Directory of Business Incubators* (Carlise, Pa.: NBIA, 1988).

F. Norling and D. Allen, "Patterns of Faculty Research Commercialization and

Contracts with Corporate Clients," paper presented at the Annual Research Conference of the Association for Public Policy Analysis and Management, University of Texas at Austin, October 30, 1986.

C. Rodenberger and J. McCray, "Start-ups from a Large University in a Small Town," in *Frontiers of Entrepreneurship Research*, K. Vesper, ed. (Wellesley, Mass.: Babson College, Center for Entrepreneurial Studies, 1981).

R. Roy and D. Shapley, *Lost in the Frontier: U.S. Science and Technology Policy Addressed* (Philadelphia: ISI Press, 1985).

J. Ryans and W. Shanklin, "Implementing a High Tech Center Strategy: The Marketing Program," in *Creating the Technopolis*, R. Smilor, G. Kozmetsky, and D. Gibson, eds. (Cambridge, Mass.: Ballinger Publishing, 1986).

R. Smith, M. Drobenstott, and L. Gibson, "The Role of Universities in Economic Development," *Economic Review* (November 1987): 3–21.

SRI International, *The Higher Education–Economic Development Connection* (Washington, D.C.: American Association of State Colleges and Universities, 1986).

D. Steinnes, "Evaluating the University's Evolving Economic Development Policy," *Economic Development Quarterly* 1, 3 (1987):214–25.

5

University Technical Innovation: Spin-offs and Patents, in Göteborg, Sweden

Douglas H. McQueen and J. T. Wallmark

During the past ten years, universities have increasingly attracted attention as sources of and inspiration for renewal of industry. Traditionally, universities have resisted this role, fearing that economic pressure and commercial interests might jeopardize the basic university values of intellectual freedom, free search for truth, and basic research into problems formulated and pursued only for the purpose of widening the frontiers of knowledge. Yet university values are always based on some ultimate benefit to mankind or society, even though remote in time and impossible to formulate in clear terms as to what the results are going to be.

Today the view that long-term basic research at universities benefits from incorporating some short-range goals is gaining momentum. One important mechanism for introducing these short-range goals is the spin-off company. Another is the production and licensing of patents on products and services developed by university employees. A third is technical consulting for outside firms and organizations. These activities can complement and enhance traditional academic research at a university.

During the past ten years Chalmers University in Sweden has actively supported the formation of spin-off companies based on the results of research carried out by members of the university staff, graduate students, and undergraduate students. Following are some advantages and disadvantages of such support.

Spin-off Advantages to Universities

1. Spin-off companies contribute to an exciting atmosphere at the university when new and important practical results are pursued and when considerable wealth can be acquired by those working on them. An exciting atmosphere is an important attribute of a university, creating many kinds of direct and indirect motivations to students and staff.

2. Spin-off company activities have a positive influence on research—basic and applied—at the university in many ways. Important needs to meet specific goals are brought to bear on the choice of alternatives in basic research. This tends to bring about relevance, immediacy, and efficiency.

3. Spin-off company activities have a positive influence on teaching at the university by introducing company activities in a realistic way—that is, the manner in which society can make use of new ideas and products. As most university graduates will spend their lives in companies, an early and realistic meeting with this environment is important.

4. Spin-off companies provide improved possibilities for undergraduate and graduate thesis work and other educational projects at the companies. Direct participation by spin-off company personnel in part-time teaching and research is very valuable. Key persons may be lost to the university when recruited by industry, but they can still participate in university research and teaching when at a spin-off company that is near campus.

5. The regional role of the university is greatly enhanced by the presence of spin-off companies. This gives the university a stronger position in interaction with regional officials.

Spin-off Disadvantages to Universities

The advantages to the university of spin-off companies should be balanced by consideration of possible disadvantages, among which are the following:

1. Spin-off companies may be perceived by the university as a threat to intellectual freedom. However, this point appears somewhat academic and exaggerated. Until tangible evidence can be found and drawbacks demonstrated, this point merits little concern except as a warning.

2. Increased size of the university through the addition of spin-off activities more or less incorporated within the university may be seen as a disadvantage. Increased size almost always contributes undesirable side issues. However, advantages seem to dominate, judged by the fact that growth is the rule.

3. The values of the university may change. A certain change of values is indeed what we are witnessing today. In view of the very desirable results obtained, it is difficult to disapprove of this. Rather, some further change—within reason—along the same line appears desirable. In reality the change in university values is still marginal.

Advantages to Spin-off Companies

1. Location of the spin-off company on campus allows it a "soft start." This can be realized by part-time employment in the company and part-time at the university. Expensive equipment can be rented from the university, saving the company the prohibitive cost of acquiring the equipment themselves. Specialized services can be purchased from the university. In addition there are good opportunities for participating in further education at the university. One condition is that the spin-off company contribute to the research at the university.

2. Spin-off companies located on campus have a favored position in recruiting—something that may be a decisive factor in a successful start-up.

3. Spin-off companies that retain contact with the parent research group have an important advantage in research backup, particularly when it comes to launching a second product. Sooner or later the product of a spin-off company will reach its end by competition, by technical progress, or by saturation of the market. The generation of a second product to supplant the first is often beyond the power of a spin-off company. University research may be of crucial importance as a possible source of this second product.

4. Frequent interaction between colleagues within the university and the spin-off company environments provides stimuli and role models and frequently expert advice of great value.

5. The prestige of the university may be of great help to a fledgling company with little credibility and financial resources.

6. Formal university support, in problem areas that spin-off companies meet early (such as patenting, financing, marketing, management, incubator housing, and company formation), is of considerable help. Formal courses are useful, but individual coaching by teachers or by consultants is even more beneficial.

Advantages to the Region

Although university research and education are usually considered in a national or international context, the influence on the region in which the university is located is more direct and easier to define. The advantages to the region of university sponsored spin-off companies are, among others:

1. The economic impact of the university on the region is increased (in Göteborg, by about 50 percent) by the annual turnover of the spin-off companies. At Chalmers the university's annual budget is about US$110 million, and the current annual turnover of the spin-off companies located in the Göteborg area is more than US$50 million. The total turnover of the Chalmers spin-off companies founded since 1964 is in excess of US$100 million. Spin-off companies subcontract production work and purchase specialized services, thus creating an infrastructure of production, service, building, and transportation. Spin-off company personnel cause an expansion of

community and private services, such as schools, health care, housing, automobiles, recreation facilities, banks, and shops.

2. Spin-off companies represent high technology and contribute to modernization of the economy of the region. In regions like Göteborg where a few large companies have traditionally dominated the economy, the addition of spin-off companies results in a more diversified and therefore more stable economy—less sensitive to failure of a single company or branch of industry, such as shipbuilding or steel.

3. Spin-off companies are dynamic and expansive, setting examples that stimulate other new companies into existence.

4. The cultural base of the community is strengthened and enlarged by the presence of additional academically trained people at the university and in its vicinity.

Spin-off Companies at Chalmers

In the autumn of 1987, the information on Chalmers spin-off companies was updated; and in addition, data were compiled on spin-off companies from two other Swedish universities with technical faculties: Linköping University, and the University of Lund (Figure 5.1). At present, Chalmers produces between 10 and 15 spin-off companies annually. The majority of these spin-offs are very small consulting or computer-related enterprises; and about 25 percent are based on new, often patented products. Experience has shown that both consulting companies and computer-related companies can grow at least as quickly as manufacturing companies. Today, two Chalmers spin-off companies are listed on the Swedish OTC stock market: one in management consulting, and one in computer engineering.

A comparison of the Chalmers data with the data from Linköping and Lund shows that Chalmers spin-off activities started about ten years earlier. Chalmers has provided greater support for spin-off formation for a longer period of time than the other Swedish universities. At Chalmers, support for new companies comes in several forms:

• Education programs. Undergraduate and graduate courses in innovation, consulting, patenting, and business formation.

Figure 5.1   Annual Formation of Spin-off Companies from Chalmers University of Technology, Linköping University of Technology, and the University of Lund, 1960–87

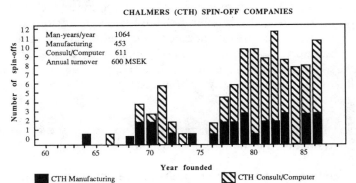

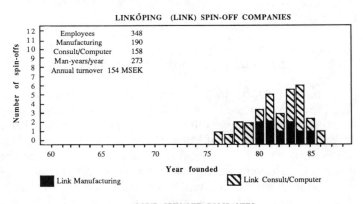

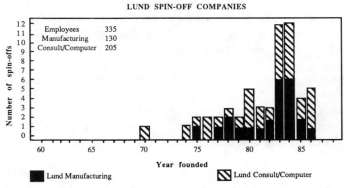

Source: The Authors

• Consulting and advising. Individual help in setting up companies, in renting space (sometimes from Chalmers), and in writing contracts.
• Financial support. Help in obtaining research contracts, bank loans, and venture capital. Chalmers started its own venture capital company for this purpose.

In the Swedish system, a spin-off company must be an entirely independent entity—most often, a corporation under Swedish law. This means that if a faculty member wishes to devote some time to a spin-off, then he or she must take an academic leave of absence or else work at the company only part-time. The latter arrangement is encouraged by Chalmers, since in the former case Chalmers no longer benefits from the faculty member's career. A certain amount of skill and control is required to keep spin-off company finances and public-sector finances separate.

Spin-off companies are concentrated in branches of industry characterized by intensive R&D, high knowledge or information content, and expansion. Figure 5.2 and Table 5.1 show what this means in numerical terms. In Figure 5.2 the amount of R&D in each branch of industry as a percentage of the value added for that branch is plotted along the X-axis, where R&D expenses are direct costs plus 50 percent for administration, capital costs, and quality control. On the Y-axis, the ratio of the value added in 1983 to that in 1972 is plotted for each industry branch. Thus in Figure 5.2 R&D-intensive, expansive branches are in the upper right; and R&D-poor, shrinking branches are in the lower left. The numbers next to some of the industry names pertain to the number of Chalmers spin-off companies, in particular those which manufacture and sell physical products. For example, there are 11 Chalmers spin-offs making electrical equipment, 15 making instruments, and three in transportation equipment. The correlation to R&D intensity is strong indeed. We also note that this industry-branch distribution of spin-off companies is much different from the industry-branch distribution of undergraduate or graduate students—for instance—or allotted university funds.

Table 5.1 illustrates another way of presenting the data, this time including all the Chalmers spin-off companies—manufacturing, computer related, and consulting. By studying Swedish stock-market reports for 1984, a ratio of market value to value added for various industry branches can be estimated. A high market value to value added ratio is indicative of a premium price paid for the shares of companies in this particular branch of industry. This in turn is generally taken as indicative of high hopes for the future of the companies in question. Table 5.1 shows the distribution of 113 Chalmers spin-off companies among various industrial

Figure 5.2   Chalmers R&D Expenses in Percent of Value Added according to Industry Branch, against Growth in the Respective Branches of Industry

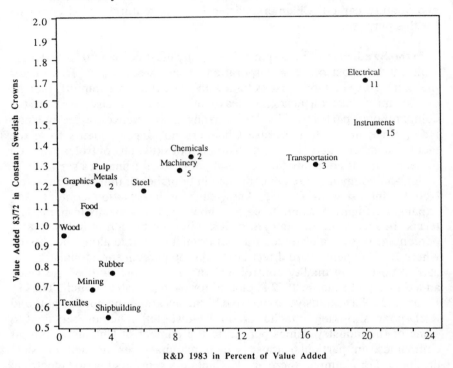

Source: For value added, Statistics Sweden SCB N 10 SM 8501 Appendix 4, and for R&D, the Royal Swedish Academy of Engineering Sciences.

Note: Value added in 1983 and 1972, in constant Swedish Crowns. R&D expenses are direct costs plus 50% for administration, capital, and quality control.

branches. With the exception of the pharmaceutical business (where Chalmers has no spin-offs) Chalmers entrepreneurs seem to choose areas where stock-market hopes are high.

Patent Activity

In Sweden, university researchers retain all rights to patents that they apply for which are based on their research. If a company or other financier requires a patent rights agreement with the researcher, then the

Table 5.1   Distribution of Chalmers Spin-off Companies with Respect to Industry Branch and Stockholm Stock-market Value as of 1984

Number of Spin-offs	Industry Branch	Market Value / Value Added
0	Pharmaceuticals	2.8
15	Instruments	2.2
31	Electronics/Computers	1.8
50	Consulting	1.4
2	Chemicals	1.1
5	Electrical Equipment	1.0
0	Rubber Products	0.8
3	Transport Equipment	0.7
4	Machinery/Equipment	0.6
3	Metal Products	0.6
Total 113		

Note: The value is divided by value added for each industry branch (based on a market sample only).

Source: The authors.

rights may be transferred. For almost all government grants, however, the individual retains the right to patents based on his or her work. This means that the researcher must also finance the patent applications and licensing agreements. In some instances, government loans can be obtained to cover these costs. Such a policy demonstrates how it is possible to work on scientific and technical inventions as a fully approved part of academic work at a Swedish university.

Figure 5.3 shows the results of a poll on patent activity at Chalmers that was carried out in 1978 and 1983 (McQueen and Wallmark 1984). The data for 1983 are incomplete. In all, 170 patent applications corresponding to 170 different inventions were recorded. Commercial agreements covering at least 100 of the patent applications have been signed with companies for exploitation of the inventions. In addition, about 25 patents have been exploited in Chalmers spin-off companies—that is, by the inventors themselves. There is an intimate connection between patent activity and the formation of spin-off companies at Chalmers. Finally, less than 50 patent applications have not led to exploitation in some form.

Innovation and Academic Achievement

Debates on the influence of patent and innovation activity on the academic work of the teaching staff of the university are common. In

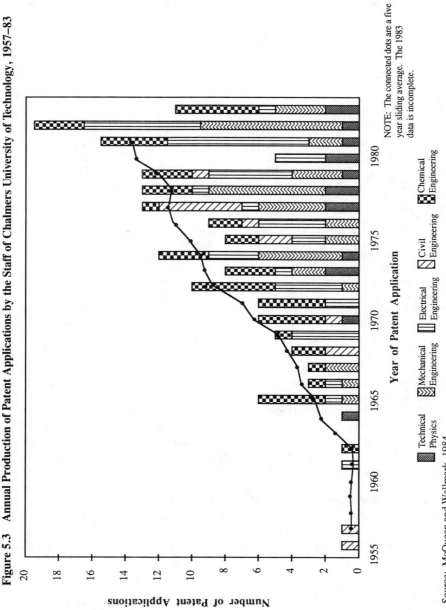

Figure 5.3   Annual Production of Patent Applications by the Staff of Chalmers University of Technology, 1957–83

NOTE: The connected dots are a five year sliding average. The 1983 data is incomplete.

Source: McQueen and Wallmark, 1984.

order to determine whether there is any correlation between patent activity and scientific achievement or production, science citations were studied. A widely accepted measure of scientific output is the number of citations to an author's work by other authors (Garfield 1979)—the theory being that this reflects the value of the author's work to others active in the field. Three five-year cumulative indexes for authors—1965–69, 1970–74, and 1975–79—were used to find the number of citations listed for each inventor with at least one patent application. Self-citations were eliminated.

Figure 5.4 shows that, among Chalmers inventors, there is a positive correlation between patent activity as reflected by number of patent applications and the same inventor's science citations. The Spearman rank correlation coefficient—based on more than 40 data points—is ($r = 0.60$) positive and large (Wallmark, McQueen, and Sedig 1988).

Since a correlation by itself does not indicate a cause and effect relationship, it is not possible to declare that patent activity aids outstanding research. However, the results do not support the argument that academic innovation activity is detrimental to the quality of academic research. In fact, the opposite is more likely to be true. That is, industrially oriented efforts leading to patent applications and spin-offs are likely to go hand in hand with outstanding academic achievement.

Conclusion

The fact that outstanding academic achievement and commercially oriented activities such as consulting, patenting, and company formation can be combined to advantage should not come as a surprise. One of the most famous scientists of modern times—Albert Einstein—had about 15 patents. Indeed, he worked in the Swiss patent office for several years. Einstein is not a special case. All of the Swedish Nobel prizewinners in physics have patents—including G. Dalen, M. Siegbahn, H. Alfven, and K. Siegbahn. Among Swedish chemists, the Nobel prizewinners A. Tiselius, G. Hevesy, H. von Euler-Chelpin, and T. Svedberg all have patents. The message is clear: Patent activity and outstanding research results do go hand in hand. It seems probable that the same positive correlations will be found between consulting activity and academic research, and between spin-off company formation and academic research; but detailed studies on these particular questions have yet to be carried out.

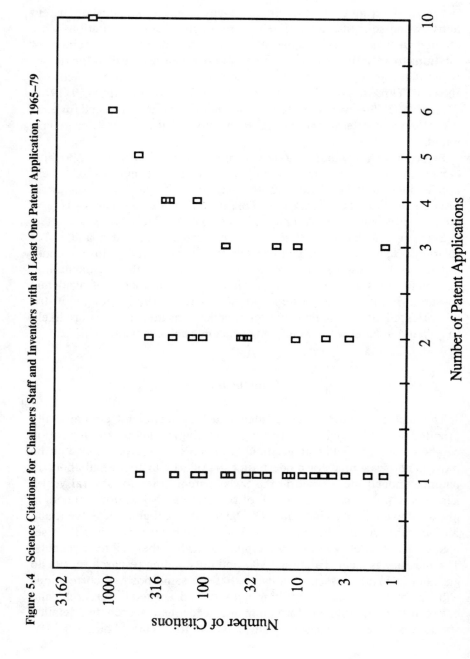

Figure 5.4 Science Citations for Chalmers Staff and Inventors with at Least One Patent Application, 1965–79

Source: Wallmark, McQueen, and Sedig, 1988.

References

E. Garfield, *Citation Indexing* (New York: Wiley, 1979).

D. H. McQueen and J. T. Wallmark, "Innovation Output and Academic Performance at Chalmers University of Technology," Omega 12 (1984):457–64.

J. T. Wallmark, D. H. McQueen and K. Sedig, "Measurement of Output from University Research," *IEEE Transactions on Engineering Management* 35 (1988):175–80.

6

A Supportive Environment for Faculty Spin-off Companies

James D. Morrison and William E. Wetzel, Jr.

There is probably no single best way to facilitate the transfer of new ideas from universities to the commercial sector. The spin-off of faculty companies is just one of several methods—but one that some universities, especially public universities, find troublesome and possibly threatening. While not without acknowledging the potential problems, we propose that, even at a small public university such as the University of New Hampshire, a supportive environment for the spin-off of faculty companies is a vastly superior environment for enlightened learning and creative endeavor than one designed to restrict faculty inventiveness and entrepreneurship. The ivory-tower notion that all university faculty should confine their activities to a narrow academic definition of teaching, research, and service is one that has outlived it usefulness—if indeed it ever had any.

We have created at the University of New Hampshire (UNH) a support system for faculty who want to establish companies, carry out applied research, develop intellectual property, or in other ways participate in the commercialization of the products of their creative energies. Elements of this support system are also available to the general public. This chapter focuses on one such element: the Venture Capital Network (VCN). VCN is a "dating service" that seeks to bring together inventors/

entrepreneurs and informal investors, who are sources of first-stage venture capital in the $50–750K range. VCN has been "franchised" to other regions of the United States and Canada.

There are five basic elements in the UNH support system for faculty inventors/entrepreneurs:

1. A university administration that encourages faculty entrepreneurship, and university policies that facilitate it;

2. The Industrial Research and Consulting Center, which provides access to university facilities and services and acts as a conflict of interest buffer and a source of advice and encouragement;

3. An innovation incubator—a physical facility where those in the process of business development can maintain an office and a prototype development area in an environment where such activity is the main enterprise;

4. A network of not-for-profit and for-profit advisory services; and

5. The not-for-profit Venture Capital Network.

University Administration and Policies

Administrators at the University of New Hampshire are openly supportive of faculty entrepreneurship. The University Research Office sponsors an Outstanding Innovator Award, given every two years to a faculty member who has added value to a creative idea through entrepreneurial effort. The award consists of a $5,000 unrestricted grant and a dinner in the award winner's honor. One of the authors of this chapter won the award for 1987–88 for his development of the Venture Capital Network. More than 250 guests attended the award dinner—including the chairman of the university system board of trustees, the chancellor of the university system, the president of the university, and state business leaders. Support of the award by university leaders demonstrates a commitment to the concept of faculty entrepreneurship.

The university's conflict of interest and intellectual property policies allow—and in fact support—faculty involvement in private-practice consulting activities and invention development. These two activities can lead to faculty entrepreneurship, which in turn can eventually stimulate the formation of spin-off companies.

Conflict of Interest Policy

The university's Conflict of Interest Policy for faculty is very loosely and generally worded. Its primary tenets are that a faculty member may not use university facilities for personal gain and cannot make a purchase decision on behalf of the university from which he or she (or a family member) benefits financially. Faculty are also constrained from usurping university business opportunities. The policy allows private-practice consulting on a 20-percent time (also known as the "one day per week") basis—provided obligations to the university are not neglected. Administration of the policy is the responsibility of department chairs and deans, with oversight by the associate vice-president for research and the vice-president for academic affairs. Faculty members report to their chairs and deans via annual reports, and also informally. The object is to make the administration generally aware of consulting relationships. There is no close or official monitoring, nor is there any formal accounting of fees or reporting of total income from such activity.

Faculty and staff (including administrators) are encouraged to participate in professional activities as a means of improving not only their own competence and prestige, but the prestige of the University System of New Hampshire (USNH) as well. While engaging in these activities, they have the obligation to avoid ethical, legal, financial, and other conflicts of interest so as to ensure that their outside activities and interests do not clash with their primary responsibilities at the institution. Therefore, USNH employees must be sensitive to the potential for conflict of interest in their professional activities. The fact that the USNH is a public institution must always be kept in mind. Conflict of interest in general terms, and more specifically as detailed in the text of the written policy, is prohibited.

Faculty may not use university-system facilities, supplies, materials, equipment, or services for their professional activities without first obtaining approval of the appropriate department chairperson, dean, or director and arranging to the total cost of such use. Such prior approval is not necessary, however, when the facilities, supplies, materials, equipment, and services are generally available to university system faculty on payment of an established fee and the fee is paid.

The name of the university system or the campuses may be used in connection with professional activities where necessary to identify the faculty member, but may not be used to imply that the university officially supports, endorses, insures, or guarantees the results of the professional activity. When the potential for confusion about official endorsement exists, a disclaimer should be used.

Faculty members who have or who reasonably anticipate having either an ownership interest, a significant executive position, or a consulting or other remunerative relationship with a prospective supplier may not participate in the recommendation of, drafting of specifications for, or the decision to purchase the goods or services involved. Faculty members who know that a member of their family (or any other person with whom they have a close personal or financial relationship) has an ownership interest or a significant executive position in a prospective supplier's company are also disqualified from participating in the purchasing of goods and services from that supplier.

When a faculty member is disqualified from participating in a procurement decision, the fact of the disqualification and the reason for it must be reported to others involved in the decision. If necessary, a substitute may take the staff member's place under procedures established by the appropriate administrative official.

Faculty members are prohibited from usurping business opportunities of the USNH if such opportunities become available to them through their employment activities with the University System of New Hampshire.

Intellectual Property Policy

The university's Intellectual Property Policy is designed to encourage invention, creativity, and entrepreneurship within the university community. It provides for the sharing of royalties with inventors and their academic units. The policy is administered by the Office of the Associate Vice-president for Research. Legal services are provided through contractual arrangements with intellectual property specialists at an area law school and nearby private-sector law firms.

When the formation of a faculty spin-off company is contemplated, the associate vice-president for research (AVPR) meets with the parties to determine if there is any involvement of existing intellectual property that has been developed using university facilities or resources (including faculty and staff time). If there is, a contractual agreement (essentially a license) is negotiated.

The AVPR has the authority to make unilateral decisions about intellectual property, unencumbered by a committee process but bound by the university's Intellectual Property Policy. That policy, like the Conflict of Interest Policy, is intended to be adaptive rather than bureaucratic. While protective of the university's rights and its intention to promote the public welfare, it allows the AVPR a considerable degree of discretion and flexibility. The university wants to see intellectual property developed and used and is not reluctant to grant exclusive licenses and facilitative terms to promote this end. A rendering of the policy as it applies to patents and copyrights is given at the end of this chapter.

The Industrial Research and Consulting Center

In 1982 UNH established an office to promote private-sector interactions and entrepreneurial activity. This Industrial Research and Consulting Center is part of the University Research Office. The associate vice-president for research delegates to the center responsibility for negotiating and administering industrial contracts and some state-agency grants and contracts. The center will also assist faculty with the negotiation and consulting agreements and the management of consulting contracts that require the use of university facilities.

The associate vice-president for research and the director of industrial research and consulting also advise and assist faculty who wish to start a company. The Industrial Research and Consulting Center is, in effect, a broker of university facilities and services. In the process of forming a faculty spin-off, it functions as a conflict of interest buffer. This role is best described by example.

An engineering professor—let's call him "Newton"—wanted to form a specialized research-and-development company. He sought the assistance of the Industrial Research and Consulting Center. The director of the center helped Professor Newton negotiate a business incubation agreement with his department and college. Laboratory and office space leases were signed. Professor Newton's company entered into contracts with the center (acting on behalf of the university) for specific research-and-development work. The center—using release-time agreements—hired Professor Newton to do the work, but took some financial and administrative control of the project out of his hands in order to avoid a conflict of interest situation. In effect, the center's director acted as principal investigator for the project as far as personnel decisions were concerned. If, for example, Professor Newton needed to hire a graduate student to do project research, then the director of industrial research and consulting hired an appropriate person over whom Newton had no "academic control"—that is, was not the professor in any of his or her classes, and was not on the thesis committee. During the period of incubation, all of the contracts for university services—including the release-time agreements—carried an indirect cost rate that reflected the "actual cost" of the arrangement to the university (including benefit costs associated with Newton's release time).

Over a three-year incubation period Professor Newton established his business, secured outside financing, and built company offices and laboratories in a neighboring community. Some recent university graduates are company employees. The university has no equity position or other financial interest in the company.

Innovation Incubator

The University of New Hampshire has entered into an agreement with Ventures Incorporated—a Manchester, New Hampshire, company—to operate a business incubator. The physical facility is located about a mile from the center of the campus. Ventures Inc. has similar agreements with several other universities in the United States and Canada.

The incubator is operated under the following arrangement. The university leases space to Ventures Inc. at a low rate per square foot. Ventures Inc. passes this rate through to participants in the incubator. Through a subsidiary, potential incubator participants are screened by an advisory board made up of representatives of Ventures Inc. and the university. The university and Ventures Inc. each obtain 5-percent equity in the participant's company and certain options to provide venture financing in return for more equity later. The university provides basic clerical services to incubator tenants and access to university facilities and services. Charges for clerical services are built into the lease fee paid by Ventures Inc. Other charges are based on a fee schedule developed by the UNH Industrial Research and Consulting Center.

During residence in the incubator (18–24 months), the participants write business plans, carry out proof-of-concept experiments, and build prototypes. Ventures Inc. delivers management, financial, and legal advice to participants through a network of mentors (successful entrepreneurs) and arrangements with professional providers.

Before the agreement with Ventures Inc., the university had considered several other offers to establish business incubators on its property.

These proposals were all essentially real estate deals. The Ventures Inc. prospectus seemed best because it did *not* include an arrangement whereby Ventures Inc. would lease space from UNH and then sublease it to participants at a higher price. Ventures Inc. makes its profit on the management, financial, and legal services and (hopefully) on its initial retained equity and later venture capital financing. Furthermore, Ventures Inc. is connected to a medium-size venture capital fund under the same management.

It was believed that faculty entrepreneurs might want to put administrative distance between their emerging ventures and the campus community. And UNH administrators decided they did not want to run the incubator. If an incubating venture should happen to fail, it is the responsibility of Ventures Inc. to close it down. The university stays at arm's length, ready to accept the entrepreneur back as a faculty member without having played a role in terminating the failed venture.

Advisory Services

Participants in an early start-up need lots of informal advice and contact with sympathetic listeners. Ventures Inc. provides some of this, but UNH decided to house in the incubator other entities with an interest in entrepreneurship and emerging companies. By doing so, the university creates an instant "critical mass" of like-minded spirits.

The staff of the Venture Capital Network, Incorporated (VCN) is one such group. Another is the staff of the Center for Venture Research (CVR), a multidisciplinary center for the study and promotion of innovation, entrepreneurship, and economic development. A third is the staff of the Small Business Development Center (SBDC) field office. Surrounded by people who thrive on entrepreneurial activity, the incubator participant finds immediate support and acceptance; and this network of support is available on a continuing, daily basis.

The Center for Venture Research (CVR) is the university's liaison with the Venture Capital Network. There is a sharing of operating and professional staff between the two organizations, and the president of VCN is also the director of CVR. CVR operating staff also function as the clerical services providers for participants in the incubator and the Small Business Development Center field office. By sharing staff, a high level of cost effectiveness is achieved.

The professional staff of CVR is drawn primarily from the faculty of the Whittemore School of Business and Economics. CVR is engaged in a variety of research projects important to emerging companies and small businesses. CVR conducts research on such subjects as the financing of small business, tax issues affecting small business, and relevant public policy issues. In addition to its role as an academic institute, CVR carries out contract research for professional associations, foundations, non-profit entities, business organizations, governmental agencies, and private-sector companies and individuals.

Venture Capital Network

One of the most critical needs of emerging companies is capital. Normally, personal funds are sufficient for only the initial steps required to launch a business; financing growth requires other resources. Substantial growth is rarely capable of being funded from retained earnings and internal cash flow alone.

Professional venture-capital fund managers generally like to make substantial investments (more than $1 or $2 million) and are most often attracted to high-technology ventures that can go public or be bought out rather quickly. Entrepreneurs can best obtain seed, start-up, and initial growth financing from informal investors.

In 1980 a study of the informal venture-capital market in New England found that a large, but rather inefficient network of funding "angels" existed. The typical deal was in the $250–500K range and in the $20–50K range per investor. The total available pool of informal venture capital was estimated to be perhaps three times as large as the professional venture-capital pool. But contact between investors and entrepreneurs was largely a matter of chance.

To make the process more efficient, the Venture Capital Network, Incorporated (VCN) was developed at the University of New Hampshire. VCN is a data base that matches the interests of informal investors with those of inventors and entrepreneurs. Like a dating service, the process begins anonymously. Potential investors review profiles of opportunities in their areas of interest (Stage 1). If any appear to be of interest, a business plan summary can be scanned for additional information (Stage 2). Continued interest results in a VCN introduction of the investor and entrepreneur (Stage 3). And that is as far as VCN goes. After the introduction, the parties are on their own.

Since November 15, 1985, VCN has listed over 200 investors and over 500 entrepreneurs. More than 11,000 Stage 1 profiles have been provided to investors; more than 3,000 Stage 2 business plan summaries have been sent; and over 1,150 Stage 3 introductions have been mediated.

Intellectual Property Policy at the University of New Hampshire

Section 1: Preamble

The promotion of research and scholarly writing is an essential function of the University. In order to foster such activities, the University shall maintain a policy for the handling of intellectual property, such as patents, copyrights, and other proprietary work, which will be generally favorable to the inventor or author.

Section 2: Intellectual Policy

The Principal Administrator responsible for Research at the University shall administer the intellectual property policy of the University.

The Principal Research Administrator shall assume the following responsibilities but may delegate operational details to Research office staff and/or knowledgeable members of the University faculty.

2A. To act in accordance with the policy here set forth.

2B. To make such recommendations to the President with respect to any changes in the intellectual property policy of the University as may, from time to time, be deemed advisable.

2C. To determine whether or not the University has an interest in any invention or discovery made by a member of the faculty or staff or by a student. Where such an interest is found to exist, the Principal Administrator for Research shall act in accordance with the policy here set forth, and, when necessary, shall advise the President and the Board of Trustees of the University what steps should be taken to protect and develop the University's interest.

2D. To see that files and records of intellectual activities are maintained and kept current in the Research Office.

2E. To submit an annual report of intellectual property activities to the President of the University.

2F. To publicize the University Intellectual Property Policy and activities.

2G. To offer information and assistance to University members concerning procedures that should be followed in order to gain adequate protection between the time of conception of an invention, discovery or copyrightable work and the processing of a formal application for a patent, copyright or proprietary agreement.

Section 3: Invention Classification

For the purpose of classification, the University shall recognize that inventions fall into the following three categories.

3A. Inventions that do not involve the significant use of funds, space, or facilities administered by the University are wholly the property of the inventor. (Significant use is construed to mean University involvement in excess of payment of salary, provisions of office and/or laboratory space, or use of library).

3B. Inventions that involve the significant use of funds, space, or facilities administered by the University, but without any obligation to others in connection with such support, are the property of the University if the Principal Research Administrator so decides.

3C. Title to inventions that result from projects sponsored by an agency outside the University shall be governed by the contract agreement between the University and the sponsoring agency, or in the absence of such contractual provisions shall become the property of the University.

The Principal Administrator for Research shall recommend to the President of the University which of these methods of basing the awards shall be followed in each case.

Section 4: Reporting of Discoveries and Inventions

University members involved in projects from which an invention or discovery is likely to arise shall keep adequate records, witnessed where necessary, and shall report promptly to the Principal Administrator for Research any inventions or discoveries whether or not the inventor believes the University has a direct interest in the invention or discovery. The property rights to such records shall depend upon the classification of the invention (Section 3).

Section 5: Patent Assistance Organizations

The University has entered into a Patent Assistance Agreement with Research Corporation. Under this agreement Research Corporation will assist the University and its inventors with the evaluation, patenting, and licensing of their inventions through assignment of their patent rights to Research Corporation.

However, the University, with the consent of an inventor, may make any other arrangements for the patenting and exploitation of an invention provided that the financial benefits to the inventor and to the University appear to be at least as favorable as those which would be afforded if the invention were administered by Research Corporation under its agreement with the University.

Section 6: Patenting of Inventions That are the Property of the University

When the question of patenting an invention or discovery is brought to the attention of the Principal Administrator for Research, he/she will take steps to insure that the invention or discovery is adequately documented. He/she will then consider whether the invention is worth patenting. In making this decision, the Principal Administrator for Research may, whenever necessary, call upon other persons, associated or not associated with the University, for technical or other advice. Furthermore, in reaching a decision the Principal Administrator for Research will consider not only the importance of the invention or discovery, but also whether or not the interests of the inventor, of the University, and of the public would best be served by a patent.

6A. If the Principal Administrator for Research decides that the interests of the public, the University, and the inventor would best be served by not filing for a patent, then a report of this decision shall be transmitted to the President of the University and the matter shall be dropped from further consideration until circumstances warrant reconsideration. If the patent application is not pursued for one of the reasons stated, the invention or discovery will remain in the University's possession and remain the exclusive property of the University.

6B. If the Principal Administrator for Research decides not to recommend a patent application but this decision is not based on public or other interest, as indicated in Section 6A, then the invention will be returned to the inventor or discoverer who shall be free to make a patent application on his or her own responsibility, if he/she so desires. If the University so waives its interest, the inventor must assume all liabilities connected with the exploitation and defense of the invention or discovery and must not use the name of the University in advertising or otherwise promoting the development, manufacture or use of the invention.

6C. If the Principal Administrator for Research decides to recommend further evaluation, he/she may ask Research Corporation or other patent development organizations for opinions about the invention or discovery. The Principal Administrator for Research will then recommend to the President of the University what other steps should be taken to protect the University's interests.

If the University decides to file a patent application, the inventor, acting through the Principal Administrator for Research, will be expected to give all reasonable assistance in preparing descriptions and illustrations of the invention.

Section 7: Income from the Sale of Patent Rights

The University will distribute license fees, royalties, and other income received from the sale of patent rights according to the following schedule:

30% of net income to the inventor(s).

30% of net income to the inventor's college or school (or program, if the inventor is not associated with a college or school).

30% of net income to a University-wide Research and Development Fund to be administered by the Principal Administrator for Research.

10% of net income to the Research Office to provide funding for intellectual property administration and development programs.

Net income is to be interpreted as that amount of money cumulatively received after deduction of expenses connected with developing, securing and maintaining the patent.

Where an invention is conceived jointly by two or more inventors, then each of the co-inventors shall share in the gross sums of money referred to above in such proportions as the joint inventors and the Principal Administrator for Research agree reflect their respective contributions.

Section 8: Patenting Inventions Not the Property of the University

The preparation of an application for a patent and the conducting of the proceedings in the Patent Office is an undertaking requiring knowledge of patent law and Patent Office practice as well as knowledge of the scientific or technical matters involved in the invention. Although an inventor may prepare his or her own application, file it in the Patent Office, and conduct the proceedings, unless he/she is familiar with these matters, difficulties may be experienced. Most inventors, therefore, employ the services of a patent attorney or patent agent to do the necessary work, which may involve a substantial sum of money. Therefore, it shall be the policy of the University through its arrangements with Research Corporation or other patent development agencies to assist University members in the matter of obtaining patents, if requested to do so and if such requests are approved by the Principal Administrator for Research. The University also will make available to University members arrangements for patent management. The procedure to be followed is essentially that outlined in Section 6. In this case, however, the equity to be shared by the inventor and the University shall be agreed upon before beginning the patent process.

Section 9: Copyright Policy

9A. Objectives

The University of New Hampshire copyright policy and associated administrative procedures enable the University to (a) encourage traditional incentives for scholarly productivity and its dissemination by publications, (b) comply with the law and respond to its contractual objectives, (c) protect the right and equities of individuals with respect to copyright matters, (d) establish principles and procedures for equitably sharing the income derived from copyrights, and (e) guard the imprimatur of the University.

9B. Copyrightable Material

Copyrightable materials include, but are not limited to, the following examples:

Books, journal articles, reports, texts, glossaries, bibliographies, study guides, laboratory manuals, syllabi, tests and proposals;

Lectures, musical or dramatic compositions, and unpublished scripts;

Photographs, films, film strips, charts, transparencies, and other visual aids;

Video and audio tapes and cassettes;

Live video and audio broadcasts;

Programmed instructional materials;

Computer programs (software);

Choreographic work and pantomimes;

Graphic and sculptural works of art;

Drawings and plastic works of a scientific or technical character;

Architectural plans and structures; and

Dress and fabric designs.

9C. Rights and Equities in Copyrightable Materials

The categories to be considered in the determination of rights and equities in copyrightable materials are: (1) individual efforts, (2) University-sponsored efforts, and (3) externally sponsored efforts.

9C.1 Individual Efforts

Copyrightable materials (including computer software), produced by faculty, staff, and students employed by the University shall be their exclusive property except when there is significant use of University personnel or facilities (excluding library collections) or University sponsorship including the Central University Research Fund or other separately budgeted research or teaching support funds. In the case of an individual effort the individual shall bear all expenses related to the production, use, protection, and licensing or sale of the copyrighted materials.

9C.2 University-sponsored Efforts

Copyrightable materials (including software) produced by faculty, staff and students employed by the University shall be copyrighted by the University when the University sponsors the effort by providing significant use of University personnel, facilities (excluding library collections), and funds (including support from the Central University Research Fund or other separately budgeted research or teaching support). For this case the University shall bear the costs of copyright protection.

9C.3 Externally Sponsored Efforts

Rights to copyrightable materials developed as a result of work supported partly or wholly by an external agency under a grant or contract shall be determined in accordance with the terms of the contract or agreement, or in the absence of such terms shall become the property of the University.

9C.4   Copyright Royalties
Net royalty income received by the University through the sale, licensing, leasing or use of the copyrightable material under 9C.2 and 9C.3 (instances in which the University has acquired an interest) will normally be shared as described in Section 7 for net patent income. In certain cases and at the request of the inventor(s), the Principal Administrator for Research may waive the royalty distribution schedule described in Section 7 so that departments and programs may receive most of the net copyright royalty income for academic program development. In these cases an administrative charge of 10% of net royalties may be levied by the Research Office.

Section 10: Administrative Procedures/Right of Appeal
The administration of the principles and policies set forth herein shall be the responsibility of the Principal Administrator for Research. In cases where rights and/or equities are in dispute the Principal Administrator for Research shall appoint an ad hoc review committee consisting of three persons. One person shall be selected by the individual(s) to be represented, one by the Principal Administrator for Research, and one by the Vice President for Academic Affairs. This committee shall recommend an agreement which shall take effect unless a further appeal is made by the individual or individuals involved, or by the Principal Administrator for Research. In the event of an appeal the Review Committee will present the case to the President of the University whose decision shall be final and binding upon all parties.

Readings

J. Freear and W. E. Wetzel, Jr., "Who Bankrolls High-tech Entrepreneurs," *Journal of Business Venturing* (1990, forthcoming).

W. E. Wetzel, Jr., "The Informal Venture Capital Market: Aspects of Scale and Market Efficiency," *Journal of Business Venturing* 2, 4 (1987).

Part III

Technology Transfer: Issues and Initiatives

7

Technology Transfer by Spin-off Companies versus Licensing

William D. Gregory and Thomas P. Sheahen

Conventional wisdom has it that the proper role of the university in commercializing technology is that of a passive, absentee licensor of patents or copyrights. Sadly, because of this point of view, many good ideas that started at a university either were never seen again after being published in a scholarly journal or else found their way into the marketplace with no benefit to the university, because of the difficulty and uncertainties of licensing.

More recently, arguments are being offered that one of the primary functions of a university is to disseminate knowledge for the good of mankind and that this function is only truly completed for certain kinds of knowledge when a process or device using the knowledge is generally available to the public. This view is implicit in the flow diagram of Figure 7.1 which shows some of the ways universities may fulfill the mission of disseminating knowledge.

A variety of economic pressures (reductions in enrollments and tuition revenues due to a lower birthrate, and the leveling off of government funding for basic research) have caused many academicians to warm to the idea of a broader role for the university in commercialization of intellectual property. Whatever one's philosophic position on the com-

**Figure 7.1   Some of the Ways a University Can Fulfill the Function of
Disseminating Knowledge**

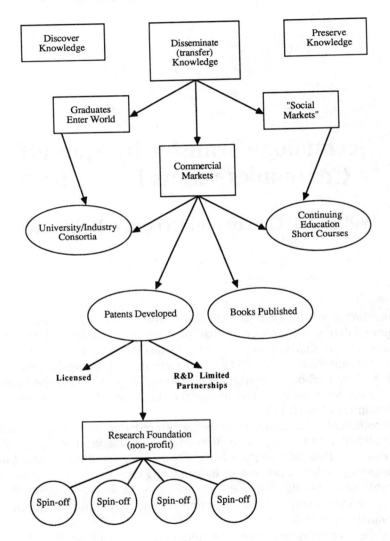

Source: The figure is found in an SRI study conducted by Sheahen and Stern (1985). The
authors have added a suggested format for a nonprofit research foundation that would
handle formation of spin-off companies to commercialize university research.

mercialization of university research, there can be no denying that the atmosphere has changed. Today it is not unheard of that universities start high-tech incubators, participate in the formation of research parks, and even take an equity position in some of the companies formed to commercialize their research results.

The authors of this chapter have participated in many projects involving technology transfer from universities, from two decidedly different perspectives: in one case, as an academician and entrepreneur; in the other, from an industrial, government, and not-for-profit institutional perspective. This chapter discusses the advantages and the pitfalls of both the traditional licensing and the spin-off company form of research commercialization. Case examples will be used to illustrate different points of view. Story contents may be slightly changed to avoid identification with any particular person, place, or institution. The intent is to share both good and bad experiences for the common good of the academic community.

Some Statistics on Licensing

There are a number of informative studies now available on the links between basic research and industrial use of the research results. One of these studies (Sheahen and Stern 1985) is particularly useful for estimating the efficacy of university patent licensing as a means of technology transfer. This study was performed by SRI International for the National Science Foundation (NSF) to estimate the value of patents issued to principal investigators on NSF engineering-research grants over a nine-year period.

Figure 7.2 is a "road map" of the SRI study, showing the steps taken to estimate the economic impact of the NSF-supported work. The SRI evaluators had available data on 4,077 principal investigators (PIs) and 722 patents issued to some (about 10 percent) of these investigators. The evaluators studied a statistically validated sample of 149 principal investigators who were issued 248 patents. Because the work was undertaken to make a connection between patent licensing and the NSF support, the evaluators first determined if the patents issued and the NSF support were linked. They then estimated the commercial value of the patents and determined the number that were successfully licensed. The final result of the study estimated the economic impact from both the patents that had been successfully licensed and the others that appeared to have some value but were not licensed at the time of the study.

Figure 7.2　Flow Diagram for the SRI Study

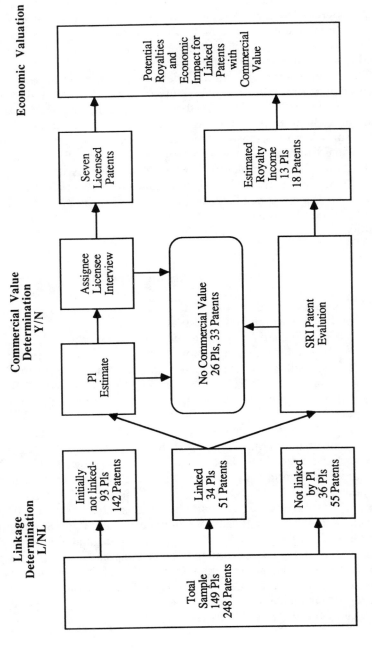

Source: Sheahen and Stern, 1985.

The results of the SRI study are normalized to a hypothetical group of 280 university researchers (investigators) to compute a "probability chain" leading from basic research to the final successful result: an income-producing license of a patent. The chain developed in Table 7.1 has the following steps: From the original group of investigators, the number of inventors is determined. For these inventors, we find the number of patents issued. For the issued patents, we then determine the number that are licensed, the number of licenses that produce any income at all, and finally the number that produce an (arbitrary) significant amount of income. For each of these steps, we compare the number of the total starting number of investigators (cumulative figures) as well as to the preceding number (incremental figures). Since we expect the chance of success to depend on the time and effort spent on the research that produced the patents, we express the cumulative figures both as percentages for the total time span of the research programs (nine years) as well as the probability per person-year of research effort. The latter figure assumes that all investigators worked in research relevant to the patented work for the entire nine-year time span, which is consistent with the assumptions we made in normalizing other figures from the SRI study.

The organization of the data in Table 7.1 is such that one can tell at a glance what the chance of success will be at any point in the process of moving from the university laboratory to industry, both overall (cumulatively) and for each added step (incrementally). The choice of steps is somewhat arbitrary, but experience suggests that they are likely the steps at which a patent licensee or licensor would pause to reflect on the wisdom of proceeding further.

The probability of producing a significant patent income is only 1.5 parts in 10,000 per person-year of research effort, which corresponds to a 0.4 percent overall success rate for the nine-year period of the study. If the entire purpose of the performed research was to make a useful result available by commercialization (which it is not), the failure rate would be 99.6 percent. An alternate view might be that, because many university research programs do not have as one of their operating goals the optimization of the technology-transfer process, the startling lack of success in this area is a self-fulfilling prophecy.

Some Statistics on Spin-off Companies

The characteristics of the spin-off companies we chose to study are listed in Table 7.2, and a probability chain of statistics similar to that presented for licensing is shown in Table 7.3.

Table 7.1   Licensing Statistics: A Nine-Year Sample

Number of Investments (man-years)	Inventors	Patents	Licenses	Any Income	"Large Income"
280 (2520)	30	50	7	5	1
Cum. Prob./ Invest.	10.7%	17.8%	2.5%	1.8%	0.4%
Cum. Prob./ Man-Year	4/1,000	7/1,000	9/10,000	7/10,000	1.5/10,000
Individual Prob.	# Inventors/ # Invest.	# Patents/ # Inventors	# Licenses/ # Patents	# Any Income/ # Licenses	# Large Income/ # Any Income
Percent	10.7%	167%	14%	71%	20%

Source: Data taken from Sheahen and Stern, 1985, and normalized. Income figures: Total income, $680,000. Income per man-year of research effort: $270.

Table 7.2 Spin-off Company Characteristics

	University Size	Location	Spin-off Product	Present Status of Spin-off
A	Large	West USA	Electronics/Cryogenics	Merged twice; on NYSE
B	Large	West USA	NDE	Merged once; on NYSE
C	Medium	East USA	Communications	Dissolved
D	Medium	East USA	Security equipment	Merged once; privately held
E	Small	Mid USA	Telecommunication Test	Privately held
F	Small	Mid USA	Test Equipment	Developing
G	Small	Mid USA	Software/Industrial Design	Developing

Source: The Authors

Table 7.2 characterizes the size of the university from which the commercialized research evolved, the geographical area of the country where the university is located, the type of business that emerged from the research, and the present status of the company. In choosing the university spin-off company combinations to use in this study, considerable effort was made to include universities in a variety of sizes (both in terms of student body and of reputation) and a variety of geographic locations. Our objective was to make most meaningful a comparison to the licensing statistics presented above. However, the spin-off data are clearly more limited than the licensing data evaluated by SRI.

Table 7.3 indicates that a group of 50 researchers were polled and that only seven spin-off companies were evaluated as having come from that group of investigators. The time frame over which these developments took place was longer by almost a factor of two than the time span of the licensing study (15 versus 9 years). The markets that the inventions in the spin-off study address are probably more limited than those available to the patents evaluated in the licensing study. For the most part, the spin-off group involved electronics—particularly computer-based systems. The one exception was a company that develops software.

We exercised a considerable amount of care in choosing institutions and researchers for the spin-off data to allow at least an elementary comparison between the licensing and spin-off data. The objective was not to choose just the success stories or major developments. Indeed, one of the companies studied was already dissolved, and most have—at one time or other—experienced serious financial and/or legal problems. If there is any bias here, it is not in favor of the spin-off data.

Despite these limitations, the findings in Table 7.3 are much more favorable than those in Table 7.1, implying that licensing is not so successful a route to commercialization as formation of spin-off companies. We believe that this is true up to a point, but a number of qualifications must be added to this conclusion that are best understood with some case examples. These examples might also help illustrate why certain things are easier to do with spin-offs than with licensing.

Internal and External Patent Management Programs

There are numerous examples of success, on a grand scale, for patent licensing programs—both those run by universities themselves as well as the many external nonprofit and for-profit patent commercialization organizations.

Table 7.3 Spin-off Company Statistics: A 15-Year Sample

Number of Investigators (Man-Years)	Entrepreneurial	Spin-off Companies	Any Income	"Large Income"
50 (750)	6	7	5	4
Cum. Prob./ Investigator	12.0%	14.0%	10.0%	4.0%
Cum. Prob./ Man-Year	16/1,000	19/1,000	13/1,000	5/1,000
Individual Prob.	Entrepreneur/ investment	Spin-offs/ entrepreneur	Any dollar/ spin-off	Large dollar/ any dollar
Percent	12.0%	117%	71%	80%

Source: Data compiled by the authors. Income figures: Total income, $896,000. Income per many-year of research effort $1,195.

The Wisconsin Alumni Research Foundation (WARF) is a bit of a cross between an internal and an external program. Founded approximately 50 years ago, WARF was set up to receive certain kinds of research funding and donations that state law prohibited being made directly to the university. At its founding, WARF was given the patent rights to the process for putting vitamin D into milk by its University of Wisconsin professor-inventor. This patent proved to be a big money-maker. Later WARF acquired the rights to a blood thinner. In appropriate doses, the blood thinner doubles as a rat poison, doubling its market. With these two income-producing patents and others that came along over the years, WARF has built $60 million worth of buildings on the Wisconsin campus and funded countless research programs.

Similar successes have been achieved by external patent management organizations such as the Research Corporation, in areas as varied as the development of the laser, materials for contact lenses, and bacteria-resistant strains of wheat. However, in almost all cases where a successful licensing has occurred via this route, the patent was a major improvement on existing technology, often leading to the foundation of a whole new industry or at least shaking up the existing industry substantially. Patent licensing organizations have learned that they must severely limit the number of invention disclosures they choose to develop and attempt to license. They have often collected a "stable" of patents in a few major market areas and achieve the most success by staying with these areas.

The downside of this story is that most disclosures made to such organizations must be rejected—in some cases, as many as 99 percent of submitted applications. While the patent managers will argue that this only represents the reality of the situation—that is, the rejected applications would never be income producers—many universities begin to have second thoughts about such advice when they see identical or similar technology adopted in the marketplace without benefit to the university research program. Clearly, there are marketable ideas that may never shake the foundations of an industry, but that nevertheless should be pursued. Licensing often is not the appropriate vehicle for such middle-of-the-road inventions.

An Example of a Direct University Attempt at Licensing

In order to be fiscally sound, a patent management program must be highly selective. This has the upside of limiting failure, but the downside of producing success only when the big winner comes along. However, such an observation does not satisfactorily answer the question of just why it is so difficult to license less-than-earthshaking inventions. After all, many rather mundane in-house inventions of even small companies find their way into the marketplace, and they yield a nice return for the developer. The two examples here and in the following section serve to illustrate another problem with licensing that answer the "why" question: The licensee–licensor relationship itself often gets in the way of successful commercialization. Stated another way, it is often difficult to get and keep the attention of a potential licensee unless an enormous reward is in the pot at the end of the development rainbow.

The first example charts the course of a licensing effort at a major East Coast university. During the course of a research effort paid for by internal university funds, a faculty member stumbled onto a very effective NDE (nondestructive evaluation) scheme. As the market did not appear large enough, and the market area not familiar enough, the external patent agency used by the university did not elect to license the technology. At the urging of the faculty member, the university pursued informal contacts through the board of directors and the alumni of the university to find an alternate route to commercialization.

The alumni and board members had very good contacts indeed. The university was put in touch with executives up to the board chairman level in about two dozen top Fortune 500 companies. In most cases, the potential product was not in line with the corporations' primary business, so such companies did not show more than a passing interest in the technology. In a few cases, however, the product did match a product line or appeared to fill a gap in the product repertoire of one of the companies contacted.

The negotiations narrowed down to three candidates: (1) a large company with a similar product; (2) a similar large company with a computer-making division that needed an end-user connection to justify the division's existence; and (3) a small company making an almost identical product that was not so versatile as the one invented at the university.

The larger company with the similar product had decided to move out of this market at about the time the university made its approach. They actually tried a reverse shuffle: Would the university like to license *their* patents?

The smaller company with the almost identical product was the best choice because of the technology match. However, they were underfunded and not inclined to offer a reasonable royalty return to the university. The university treasurer—accustomed to working with executives from major corporations and major resources—kept dreaming of a linkup with a larger firm and turned down the offer of the small company.

The second large candidate firm had to make a decision faster than the university could satisfy the items on their company checklist. The computer division was going down fast, and the firm was not going to jump out of the frying pan into the fire without a thorough "due diligence" investigation of the proffered technology. Their questions boiled down to two major concerns: (1) They wanted to see a prototype that they could show to potential customers; and (2) they wanted to see convincing market figures and market strategy, since they have never marketed in this end-user sector before. Ultimately, they decided to cut their losses and shut down the computer division.

In their final conversation with the university, however, the representatives of this firm offered a suggestion on how to go about peddling the invention. They thought it would be easier to buy a small company that had gone through the preliminary steps of building and marketing on a small scale, rather than merely licensing a paper patent. They even pointed out that they would have gladly paid much more than the university wanted for the patent rights just to have their questions answered and some of the risk removed by the trials of such a spin-off company. In fact, they pointed out that they would have chosen to keep the computer division if the university had structured their intellectual property management this way.

Armed with this advice, the inventor and some alumni put together a small for-profit corporation and leased a machine embodying the principles of the invention to a government agency. The agency used the machine in a test program for about one year and gave the technology high marks. They also permitted the university to bring interested parties to view the machine in operation. One of these visits ultimately resulted in the little spin-off company being purchased by a modest, well-heeled local firm active in another corner of the electronics business.

At this point, it might seem that this is a spin-off story and not a tale about licensing. Not so. The spin-off was, in this case, just a shell and had only the one pilot project and the intellectual property as assets. The negotiations took the form of a licensing, with the greatest concern of the new owners being on the strength of the patent applications. They did hire some university personnel who had been paid through external contract funds, and they did cover the salary of the professor-inventor for the period of time that he spent on leave at their plant helping to develop the technology further. However—and this is the most important point about the arrangement—control of the product's further development was entirely in the hands of the new owners, and the university (at their own choosing) had no equity position or voting control in the entity whatsoever. Their only connection to the project at that point—legally and financially—was the patent license. But the story goes on.

The new licensee was a privately held company, with the majority of the stock of the company owned by the executives running it. Early on in the relationship with the university, it became clear that these executives did not get along. The "angel" for the licensed technology was the company president, who was in the midst of a feud with his fellow executives. Within nine months of licensing the technology, the president was bought out by the company and they parted ways. This left the licensing project in limbo, with the funds for moving to market now drastically slashed by the executives who remained since they saw the project as one of the last vestiges of the fellow they had worked so mightily to get rid of. Within six months of the president's departure, the company and the university had found another licensee.

The newest licensee paid well for the advances that had been made while the previous company had the technology; but since the university had no equity position in that firm, they received none of the benefit. On the positive side, some litigation developed out of the sale of the assets contingent to the new licensing, which the university in the first instance avoided by being so loosely connected to the sale. And of course, the university once more had a licensee for its patents. This final licensee was the largest of the firms that had been connected with this technology.

The last licensee performed in much the same way as the previous one. The engineers at this firm developed a "not invented here" attitude about the technology and spent a good deal of the company's money trying to prove that it was not what it was cracked up to be. When they finished the prototype development—and begrudgingly admitted that the machines did indeed do what the university and the inventor claimed they would—a new factor came into play. The reason the firm had licensed the patents in the first place was that their business was growing at the rate of only about 5 percent per year and they (incorrectly) assumed that their traditional markets were drying up. For reasons entirely unrelated to the licensed technology, they found themselves a few years later with a 40-percent growth rate and a business so booming they could barely fill orders for their traditional product line, let alone get up to speed in a new marketing area. So, for probably the last time, the university's patents were unused; they are in this status at the present time.

An Example from a Major Corporation

It would be tempting to blame the problems outlined in the example above on the lack of expertise in a university setting for this kind of transaction. To be sure, the broader question of locating the correct talent pool to either license or spin off technology is one of the most important considerations involved in deciding which option to choose. However, our next example was chosen specifically to show that there are limits on the licensing relationship that even sophisticated business expertise cannot circumvent.

The invention in question was made by a scientist at one of the largest and most respected industrial laboratories in the United States. The scientist himself had received the highest of honors for his early work and was permitted by the company to choose new directions for his research as he felt appropriate. In the course of this freewheeling research program the scientist developed an extremely sensitive and inexpensive technique for diagnosing bacterial infections—sort of a "litmus test" that could be performed quickly in a doctor's office without resort to time-consuming and expensive laboratory analysis. The scientist received yet further awards for this success in this new endeavor, and his company decided to make some effort at marketing the test kits.

Even though the company in question is quite large, their main business is in the electronics industry, so they quickly decided that the method for marketing this technology would have to be different than their standard methods. The company already had a division that handled just such eventualities; its job was to choose from among their viable commercialization options—from a new venture set up internally (now generally called "intrapreneuring"), to outright licensing to an outside firm. They chose to license the technology in the pharmaceutical industry.

In short order, the company patent-license manager and attorneys found a major drug company that was willing to pick up the license. The electronics company signed a very favorable agreement with the drug firm, and development of the market-prototype diagnostic kit was passed to a group in the drug company laboratories.

Almost immediately there were problems with the development group. The first sign was that members began asking for transfers to other parts of the drug company; then others suddenly found other jobs and simply resigned. The problem appeared to stem from personality clashes between the group leader and his staff. Before the drug company could bring the problem under control, most of the staff either transferred or resigned. Then the group leader suddenly quit. After nearly a year under license, the project was no further ahead than when it started, and the drug company was faced with the prospect of forming a new team to start over on the development. After consultations, the electronics company and the drug firm decided that it would be better to cancel the license agreement. They did so, with the drug company paying a cancellation penalty that had been built into the agreement.

The licensing division of the electronics company quickly located a second drug company willing to sign the license agreement. However, within a few weeks of the signing, this drug company bought a hospital supply company to expand its base in the health-care field. The drug company had made the assumption that the new business would be similar in its operating characteristics to the drug business, so no special provisions were made for managing the new effort prior to the purchase. The hospital supply corporation was doing well, and everyone expected a smooth transition in management.

In fact, the drug company found that the hospital supply business was quite different from its own business and that it would have to send many of its top and middle management to the new acquisition to keep it afloat. Six months after the signing of the licensing agreement, the electronics company attorney made a routine telephone call to the drug company to see how the development of the diagnostic kits was going. The drug company executive who took the call was new to his job because of the management shuffling caused by the purchase of the hospital supply company; he asked for a little time to get up to speed on the project. When he finally returned the attorney's call, the executive had to admit that the project had been entirely forgotten in the chaos resulting from the new acquisition. Once again, the electronics company negotiated a penalty payment and cancelled the second license.

It was not so easy to find yet another licensee; but eventually persistence paid off, and a third interested drug company was identified and given some laboratory samples prepared by the electronics company scientist to study. Just one week before the formal signing of the third license agreement, the electronics company received a cryptic telephone call from this newest prospect, saying—in effect—that they had tested the samples sent to them, found that they did not work, and would not be signing the license agreement. By contacting friends and acquaintances in the drug industry, the electronics firm was able to ascertain that the samples had not been shipped properly and had been rendered useless by chemical contamination from the packing material. Even though the license manager pointed out to the drug firm that they had not followed the inventor's recommendations in shipping the materials to their test sites, the drug company adamantly refused to consider the project any further.

This story does have a happy ending, however. At the point in time when the negotiations with the third licensee had fallen through, a university located near the electronics company began a pilot incubator and research-park project. Perhaps sensing that they had little to lose at this point, the company entered into an agreement with a fledgling firm in the new incubator. This time they had the full attention of the smaller firm's management, since the licensed product would eventually be the mainstay of their business. In fact, the small firm was the "star" of the incubator, graduating to the industrial park in record time. After nearly a decade of trying, the electronics company had finally found a way to get the technology into the marketplace.

Spinning a Tale of Spin-off Companies

Below are some examples of the difficulties that await the unsuspecting dreamer who chooses to use the spin-off route in commercializing university research. Rather than specific cases, a grand average of the information is used to describe the proverbial "typical" case. There will be one assumption made in this story: It will be assumed that the university wishing to spin off the technology has no formal mechanism to do so, which until recently has certainly been the typical situation.

"Professor X" at "University Y" visits the intellectual property manager of the university with some exciting news. He and his students have just reinvented the wheel, and he knows a thousand places where these new gadgets can be sold. It's sure to make the university and the inventors a fortune. After checking with the external patent licensing institution used by the university and finding no interest in picking up its option to commercialize this invention, the university official agrees to license the technology to the inventor and some of his students, who will form a company to develop and market the device or process. The university stipulates, however, that this is to be strictly an off-campus project and that the university wants no legal encumbrances with the new firm, except to receive a fine royalty.

The project proceeds apace, with the professor finding a small amount of local funding and a friendly businessman with university ties who will help the project along with some funds and a lot of his expertise. Who will do the developing if the professor cannot put full time on the project outside the university? Why, no better choice than the students who were working on the research all along. This gives the university a warm feeling, since now it can say it has not only graduated the students but also cooperated in producing jobs for them.

The typical stories diverge at this point. Some of the little companies take the high road to success; but many more have a more dismal future, eventually leading to failure of the company because of inadequacies present at the very beginning of the project. The key to the difference between success and failure of these spin-offs is usually not the value of the technology or the market it has available. If the invention is not the earthshaker that the inventor thinks it is (and it usually is not), this should only mean that the company size and gross revenues will not be so large as might have originally been anticipated. Rather, the resources of the company—specifically, the talents possessed by the principals and the funds available to them for the venture—will be the key factors determining success or failure.

A bit more elaboration is necessary here. While the professor and his students may be brilliant researchers, they may not have experience in development directed toward making production-ready machines or processes, even just to the prototype stage. This will be particularly true of the student assistants, for whom this project may be their first entry into the job market.

Then there is the problem that a host of business decisions and choices must be made by someone associated with the venture. A common failing is that the inventors are called on to make these choices and they may have even less business talent than they do production development capability. Even for those faculty and students who are wise enough to recognize their limitations, there may be no good alternative to running the company themselves. If they do find some local business talent, they often become the prey of the unscrupulous exploiter, or they may find out (often too late) that people who have been in other, nonstart-up businesses are just as ill prepared as they are to run a new company.

Assuming for the moment that these problems are circumvented, the little company will eventually need full-time personnel and—most importantly—a good shot of financing, to really prosper. While the company might begin in someone's garage, at some point it will have to start spending money to make money. This puts the professor–student team into yet another area where they are likely to have no experience. This can be the most serious problem of all, since plenty of sharks swim in these waters.

Finally, even if all of these difficulties are surmounted, the entrepreneurs will eventually be introduced to the newest U.S. indoor sport called "sue your neighbor." It has become a way of life in our business and social interactions to march into the courtroom for any real or imagined grievance. No matter how small the matter, the judgments asked for can be mind boggling. Where the first suit will come from is almost a matter of random selection, but often it is a disgruntled employee, a discontent stockholder, or a customer who doesn't get the product he wanted or get it on time. In these matters, it does not help that the company personnel are young, with expectations that are unrealistic. Often, being so near the university, the student employees never quite make a transition to the working world. The time-honored student tradition of complaining about everything they don't like in university life transfers to many perceived problems that the more sophisticated entrepreneur-employee learns to work around in a constructive way.

Moreover, it is important for the university to realize that—even though it thought it had isolated itself from these kinds of problems—by virtue of being the owner of the patents, it too can be dragged into litigation. It has happened.

Most of these start-up shortcomings could be avoided—you will notice—if the university would take a more active part in the spin-off process by creating a structure and mechanisms that allow the development to take place in an orderly fashion with the proper talents and resources. Strangely enough, the best way for the university to protect itself is to become more closely involved and not isolated from the commercialization process.

Discussion and Conclusions

The statistics for success of licensing ventures is low. Because of this, patent management institutions generally hedge their bets and agree to manage the licensing of only those inventions with a more or less obviously large market that will produce substantial royalty revenues. Even when a license agreement is successfully negotiated, the project still has a substantial chance of not succeeding. The conventional wisdom that this is due to some lack in the technology is not necessarily correct. What does happen in many cases is that the licensed technology becomes somewhat of a side issue for the licensee. The cases used for illustration point out that other problems with the licensee's main business can easily distract the licensee from diligent pursuit of the commercialization. While this can be accounted for somewhat by putting terms in the licensing contract that extract penalties for nonperformance, the only further recourse is often a relicensing. For these reasons, we believe that the licensing statistics result from the loosely binding character of the relationship between licensor and licensee. The situation is not likely to be improved when the university adopts a different strategy that continues to use only licensing as a tool for commercialization. On the positive side, the larger patent management organizations have the experience and the resources to properly manage major inventions, and they can absorb almost all of the risk involved for the university in these special cases.

The statistics for non–university supported spin-off companies seem to be better than those for licensing. In particular, the spin-off does not require the large income potential that is necessary in licensing operations to make the overall scheme pay. As a result, the spin-off can produce income from more modest inventions. The problems with spin-offs are often related to the usual problems of running a business, and not to the inherent value of the technology—which is similar to the licensing experience. One serious concern is that the characteristics of the university often carry over to its spin-off companies—particularly those that use the same personnel who were originally involved in the invention's development within the academic environment. To be most successful and at the same time lower the risks for the university, a formal structure for handling spin-offs should be developed by those universities interested in using this mechanism for technology transfer. This structure should include such features as the ability to obtain financing, to build prototype systems and probe test markets, and to manage the fledgling company in an efficient businesslike fashion.

We would suggest, then, the following rule of thumb for choosing between licensing and spin-offs: If it appears that the invention in question will be a major innovation, use one of the many good external patent-management organizations. If the invention shows promise but is more modest, consider using a spin-off. When the spin-off is chosen, we would suggest setting up an organizational structure such as a not-for-profit research foundation that has the personnel to develop and manage many such activities simultaneously. The foundation, in turn, could form for-profit spin-off corporations as inventions mature into marketable prototype devices or processes.

Readings and References

R. S. Campbell and F. O. Levine, "Technology Indicators Based on Patent Data: Three Case Studies," submitted under NSF grant PRA 78-20321 by Battelle Pacific Northwest Research Laboratories, May 1984.

R. M. Cyert, "Establishing University–Industry Joint Ventures," *Research Management* 28, 1 (January/February 1985): 27–28.

IITRI (IIT Research Institute), *Technology in Retrospect and Critical Events in Science (TRACES)*, vol. 1, final report prepared for the National Science Foundation, December 15, 1968.

W. Marcy and B. M. Kosloski, "Study of Patents Resulting from the NSF Chemistry Program," Research Corporation report prepared for the National Science Foundation, September 1982.

NCURA (National Council of University Research Administrators), "Private Sector/University Cooperation: Making It Work," proceedings of national conference held in Dallas, September 4–7, 1984.

R. P. Sheahen and R. L. Stern, "NSF Engineering Program Evaluation: Patent Study," final report of SRI International project no. 6432, prepared for the National Science Foundation, February 1985.

8

Promoting University Spin-offs through Equity Participation

Meg Wilson and Stephen Szygenda

Technology transfer has usually meant the dissemination of information through seminars, classroom teaching, conferences, and publications. More recently, commercialization of research (from business or from universities) and aggressive application of existing technology to new uses have gained attention because of the staggering U.S. trade deficit, the loss of manufacturing jobs to offshore locations, and the loss of entire markets to foreign competition. This national economic challenge has led to design and implementation of new strategies, novel tools, and new programs to support active technology transfer.

Commercialization of university research is one of the most important and controversial approaches to technology transfer. Some believe that the role of a university should be restricted to education and research, that doing business or promoting technology is not an appropriate role for the university, and that a supporting role in economic development is incompatible with the mission of education (Reams 1986).

However, there are new mandates from Congress and state legislatures to promote technology transfer as a way of rebuilding the states' national competitiveness, of supporting small business development, and of re-couping—in a more direct way—something from the states' massive

153

investment in education and research. Many policymakers and business leaders now consider universities to be fellow players in the economic development arena. New university-based structures have been created over the past few years to provide institutional advocates for transfer. Their function is to be active in seeking transfer opportunities and to market the universities' research results (Bremer 1985).

These advocates—technology commercialization programs—have the traditional option of licensing university patents and know-how, or the newer option of assisting in the spin-off of a company to implement university research results. The ability of a university to own equity in the ventures that it spawns is very important to effective technology transfer. This chapter will explore the advantages and challenges of university equity participation in its spin-offs.

Tools of the Trade

The key to successful technology transfer—including the creation of spin-off companies—is the availability of a wide range of tools, clear university policies, and a supportive capital community. At the University of Texas at Austin, the Center for Technology Development and Transfer (CTDT) was created by state legislation in 1985 and granted very broad authority to support the economic growth of the state and the growth of high-technology industry by submitting to the university's board of regents

> agreements with individuals, corporations, partnerships, associations, and state and federal agencies for funding the discovery, development and commercialization of new products, technology, and scientific information. . . . The board may, either through the center or through one or more corporations incorporated by the board:
> (1) own and license rights to products, technology, and scientific information;
> (2) own shares in corporations engaged in the development, manufacture, or marketing of products, technology, or scientific information under a license from the board, the center or a corporation owned or controlled by the board;
> (3) participate as the general partner or as a limited partner, either directly or through a subsidiary corporation formed for that purpose, in limited partnerships, general partnerships, or joint ventures engaged in the development, manufacture, or marketing of products, technology or

scientific information under a license from the board, the center, or a corporation owned or controlled by the board.[1]

In 1983, Texas A&M had been granted authority to establish the Institute for Venture in New Technology (INVENT), with limited ability to own shares in ventures it spun off or assisted in (since the university primarily assisted existing small businesses).[2] CTDT understood from the experiences of INVENT that aiding small businesses to develop and/or transfer their technology was a difficult and labor-intensive effort. Thus CTDT, funded internally by the College of Engineering at UT-Austin, chose to focus its efforts wholly on commercializing university research. CTDT became fully operational in January 1987 after the board of regents approved policies governing the center's operation.

In 1987, the Texas legislature passed a farsighted package of science-and-technology initiatives. The most important of these was House Bill (HB) 1402: the Equity Ownership Bill. That bill was passed to allow all Texas public universities to participate in equity from university spin-offs and to clear up a principal concern regarding conflict of interest. The need for the legislation became apparent after the UT Health Science Center in Houston agonized over a particular spin-off opportunity.

The UT System Board of Regents had approved transfer of a medical technology from the UT Health Science Center to a start-up company. In summary, the inventor had attracted the attention of some venture capitalists (VCs) who wanted to set up a company. The Health Science Center was eager to make the deal. However, the VCs insisted that the inventor be cut in for a share of the company. To make this more palatable to the university, the VCs offered a share of equity to the Health Science Center. The UT System was unsure at that time whether either one of them—the inventor or the university itself—could accept the equity. The university was concerned about the inventor's continuing role as a professor and researcher, but even more concerned about the institution's legal position in holding equity. Would this be construed as a conflict of interest under state law? The VCs insisted on the equity issue because they wanted the inventor to have a vested interest in the success of the company. They needed to tie the research talent into the venture; otherwise, they would not have the credibility to pull in other investors, nor the confidence to stay in it themselves. After much legal soul-searching, the UT System did finally decide that it would approve the deal.[3]

If CTDT had been around when that venture was initiated, the deal could have been negotiated under the center's authority, but that would not have solved the question of the inventor's ownership of equity in the spin-off. It took the Equity Ownership Bill to do that.

The law—Section 51.912 of the Education Code—states that it is not a violation of any

> law of the State of Texas for:
> (1) an employee of a university system or an institution of higher educa-tion . . . who conceives, creates, discovers, invents, or develops intellectual property, to own or to be awarded any amount of equity interest or participation in, or, if approved by the institutional govern-ing board, to serve as a member of the board of directors or other governing board or an officer or an employee of, a business entity that has an agreement with the state or a political subdivision of the state relating to the research, development licensing, or exploitation of the intellectual property; or
> (2) an individual, at the request and on behalf of a university . . . to serve as a member of the board of directors . . . of a business entity . . . in which the university system . . . has an ownership interest.[4]

By legislative fiat, then, this law took the issue of conflict of interest out of legal consideration and strengthened the ability of universities to commercialize university research. Thus, CTDT and the other technol-ogy-transfer advocates in Texas have new and strong tools to implement their technology-transfer projects.

A companion bill—HB 1401—required every public university to estab-lish minimum-standard intellectual property policies by January 1988. These policies have to address equity ownership, faculty participation in company management, and patenting.[5]

Equity: Benefits and Uses

As indicated in the circumstances cited above, the venture capitalists wanted the inventor to have an equity participation in the start-up medical company. The demand of investors that the inventor have an incentive for remaining available and interested in developing the technology is one main reason to allow for equity ownership. A start-up company is a risky venture under any circumstance. When new science or technology is involved, the intellectual resources must be a certainty, to attract investor capital. And, as mentioned earlier, a supportive capital community is the third key to effective technology commercialization.

Besides financially engaging the inventor, however, there are other reasons to structure a transfer effort through a start-up company.

1. From an economic development standpoint, small businesses are the innovators and job creators in the U.S. economy. We are all for good statistics on the number of companies started and jobs created. It is rewarding to see a new company start, grow, and succeed, especially when that happens close to home.

2. Often a new technology requires substantial development work. Large companies may not want to license a product or process that requires nurturing before it will ever reach its maximum market potential. A few people working in a start-up company can concentrate on the technology and develop it in an evolutionary manner. The costs can then be kept very low. Start-ups often achieve amazing results with little initial investment, especially when there is a close tie back to the university.

3. The inventor (whether faculty member, staff, or student) may be reluctant to license his or her research results to a large (impersonal) company, lose control over it, and be unable to shepherd its development. A spin-off company is usually started near the university—offering proximity, access, future collaboration, and a way to keep the inventor involved.

4. Venture investors generally will not get involved in a traditional license agreement. If development research is called for, they will not fund the early work without a clear strategy for getting a high rate of return on their investment.

5. A company can be interested in licensing a university's patent; but if it does not fit their existing corporate mission, they may rather want to set up a subsidiary or help form a new company. The parent company does not want to bear the burden of paying royalties without receiving something immediately tangible for their investment. What they have to offer in return for the intellectual property rights is equity.

6. In some cases, a start-up may be the only way of supporting a technology that requires nurturing. A start-up company cannot afford the cash-flow drain of even minimum license fees and/or

royalty fees. What they can afford is to trade their equity for a license. The implications of this are that a very early technology or a very risky line of investigation may only be worthwhile on a long-run gamble. Equity in lieu of cash payments may be the best method of transferring a technology in this situation.

7. Equity provides the potential for the greatest return to the university (and the inventor). If the new technology is superseded in another two to five years, there will be little or no capture of royalties from it. Through equity holdings, however, the potential gain to the university grows as the company grows. Hopefully the company will keep expanding by diversifying its technologies and business strategies. This may be the best argument for promoting university participation in the equity of a start-up.

Equity: Cautions and Alternatives

The ability of a public university to own equity in spin-off ventures is unusual. Many observers have indicated surprise at the progressive policies in Texas and have wondered how a public institution—a non-profit organization—can participate in business ventures. As far as Texas is concerned, there are several responses.

The UT System manages assets of more than $2 billion through its Permanent University Fund. Any equity acquired through an agreement with the system will be added to that immense asset base—added to the fund's portfolio. Most public universities have an investment and/or endowment management structure that parallels the Permanent University Fund. Some states have constitutional provisions that restrict a state agency from doing anything to profit an individual business. There are sufficient legal precedents to indicate that, while universities need to be careful in their affairs, technology transfer activities do not fall under these provisions.

Many scholars, administrators, and policymakers are concerned that, if universities become too business oriented, then free inquiry will be disrupted. Active patenting and licensing seem to be the outer bounds beyond which some people feel no university should ever go. During its 1982 hearings on this controversy, the House Committee on Science and Technology referred to "conflicts of commitment" (Reams 1986, 70). During those hearings, critics such as Harvard President Derek Bok warned against university equity interests of all types.

The Texas response to all this is threefold: First, the state legislature wants to see tangible economic results from its substantial investment in university research. They see closer university–business collaboration as a way to benefit all parties. Second, a primary goal of the Equity Ownership Bill is to provide incentives for the faculty to stay at the university. Experience has shown that, when policies and procedures are not in place to help professors move an innovation on out to the market, they will often take it on out themselves. The university then loses a valuable resource. If—instead—they can get help from the institution in commercializing their research and then reap its financial rewards, they have a strong incentive to remain at the university. Third, only those faculty members who want to transfer their technology are involved. The system is totally voluntary. A professor may prefer to publish research results and never pursue a commercial venture. That's fine. Many feel that way. But for those who want to see their work through beyond the research stage, alternatives now exist.

There are clearly cases where a traditional license with royalty makes the most sense. Some technologies have too small a market to sustain a spin-off company. Some require such a high initial investment or extensive marketing infrastructure that only an established company could take it on. And there is also the matter of timing: It takes time to set up a company properly. If the concern is for a rapid transfer of the technology—to hit a market window, beat a competitor to the market, or provide immediate and short-term return to the university—then licensing to an existing company is a good strategy.

Universities should never underestimate the amount of time and care it takes to start a business. If the right resources do not exist, it may be nearly impossible to get a company off the ground. And part of this relates to the background of the institutional advocates. There are too few people with strong business backgrounds who are equally comfortable in an academic environment. Advocates are most successful when they can bridge the language and culture of universities and business. And finally, universities have to accept the fact that there will be failures. Certain deals look great on the front end but not so good after several rounds of negotiation. It is okay to say no to some deals. Then too, the statistics for start-up companies cannot be overlooked. Their failure rate is high. When universities are equity participants, they must realize that they are playing the averages. But just one or two winners can make it all worthwhile.

Many of these principles may seem less abstract in light of a case study that can highlight them and explore their implications.

Nova Automation

CTDT's first completed project was the transfer of a CAD-driven, laser-sintering prototype production process. The research had been done by a graduate student in mechanical engineering. A lecturer in the College of Engineering who was familiar with the student's work spoke with the president (Mr. "X") of Nova Graphics about this technology in the fall of 1986. President X was very interested in the potential of the system and arranged to visit with the student. After the visit, President X contacted the director of the newly formed Center for Technology Development and Transfer, to see if CTDT was set up to do things such as help him get the rights to the laser system. The answer was yes.

President X had an independent consultant perform a preliminary market analysis to indicate the potential for the new prototyping process. The report was very positive. At that point, the real work started. President X wanted to get worldwide exclusive rights to the laser prototyping system. UT had already approved the invention for patenting, and an application had been filed. In the course of two preliminary meetings with the principals, CTDT outlined the options open for transferring the technology.

Nova Graphics then brought back a proposal to CTDT, giving UT a substantial minority position in a new company called "Nova Automation" that would be started to commercialize the prototyping system. Nova Automation would be responsible for raising the development capital necessary to upgrade the prototype of the laser system, developing the business plan, and raising the follow-up capital to launch the business with a commercial version of the laser prototyping system. Dilution of equity would be possible—spread equally among all participants.

The basic proposal met the approval of the College of Engineering and was forwarded to the campus counsel and the vice-president's office. There it was learned that UT–Austin had never before executed an equity-based license. The administration's concerns did not relate to the structuring and development of Nova Automation. Their concerns were campus concerns: Was the inventor properly protected? Did Nova Automation want to contract with UT for further research? And if so, how would that funding be structured? Would Nova Automation pay for UT's patenting costs? Would Nova agree not to use the university's name in any of their advertising? CTDT turned its attention to the UT-oriented concerns. Those concerns met, the proposal was then forwarded to the UT System counsel.

The university office came back with an okay on the proposal if Nova

Automation would agree to pay a substantial royalty in addition to giving UT the equity. President X said this was not possible. He argued that the university obviously did not understand the implications of the royalty: It would impose a serious cash cost that would be an onerous burden for a start-up company. He also considered the royalty demand unreasonable in light of the generous offer of equity.

From the university system's perspective, they could not be certain that the equity would ever provide a return, and they were trying to get some up-front income from the deal. The draft agreement stipulated that Nova would reimburse UT for all legal fees. The return that UT wanted would be split with the inventor—providing an immediate reward to him, covering any unreimbursed legal expenses, and going toward support of research at UT–Austin. In tight budget times, these short-term concerns are understandable.

CTDT served as the intermediary in several rounds of negotiations that centered on the balance between equity and royalty (since UT insisted on some royalty) and on the responsibility for foreign patent filing costs. When the agreement seemed to be just about final, one last equity-based problem arose. Nova Graphics had decided not to set up Nova Automation as a subsidiary. Instead, they would be a stockholder and provide start-up support to Nova Automation. Another round of negotiations were required to assure UT that its equity position would not be unduly diluted before the company had any real value. With that accomplished, the board of regents approved the venture.

Nova Automation went on to win an SBIR grant from the National Science Foundation to develop the prototyping system. It now has a number of potential investors evaluating the situation. All indications are that it will raise the necessary development capital and have a very important technology on the market within two years.

Conclusion

Texas may be viewed as a laboratory for technology commercialization strategies. Equity participation by universities in their own spin-off companies is viewed as a very positive step by many—policymakers, administrators, investors, faculty, students, and businesses. There are already a handful of such equity-based companies, struggling to be successful. In two to three more years, UT will be able to conduct a more retrospective analysis of this experiment. But for now, the technology-transfer advocates are eagerly applying the new tools for their trade and trying to benefit both the state's universities and the economy.

Notes

1. Vernon's Civil Statutes of Texas, Education Code, Section 65.45.
2. Vernon's Civil Statutes, Education Code, Section 88.300–303.
3. Martin Sutter, "The Role of the Venture Capitalist in University Technology Commercialization," technology-transfer workshop presentation at the IEEE/ACM Fall Joint Computer Conference, InfoMart, Dallas, October 27, 1987.
4. Vernon's Civil Statutes, Education Code, Section 51.912.
5. Vernon's Civil Statutes, Education Code, Section 51.680.

Readings and References

Pier A. Abetti, Christopher W. LeMaistre, Raymond W. Smilor, and William A. Wallace, "The Roles of Industry, Small Business Entrepreneurship, Venture Capital, and Universities," in *Technological Innovation and Economic Growth* (Austin: IC² Institute, University of Texas, 1987).

Stephen Atkinson, "University–Industry Research Agreements: Major Negotiation Issues," *Journal of the Society of Research Administrators* (hereinafter referred to as *SRA Journal*) (Fall 1985).

Donald R. Baldwin, "Technology Transfer at the University of Washington," *SRA Journal* (Spring 1986).

Battelle Institute, "Universities and High Technology Development," report presented to Southern Regional Education Board at Atlanta, June 1983.

James Botkin, D. Dimancescu, and R. State, *Global Stakes* (Cambridge, Mass.: Ballinger Publishing, 1986).

Howard W. Bremer, "Research Applications and Technology Transfer," *SRA Journal* 17, 2 (Fall 1985):58.

Theodore Brown, "University–Industry Relations: Is There a Conflict?" *SRA Journal* 17, 2 (Fall 1985).

Herbert B. Chermside, "Some Ethical Conflicts Affecting University Patent Administration," *SRA Journal,* pt. 1 (Winter 1985) and pt. 2, (Spring, 1986).

Bob G. Davis and Jan K. Simpson, "A Simple and Economical Approach to Developing a Technology Transfer Program," *SRA Journal* (Fall 1987).

Jerome Doutriaux, "Growth Patterns of Academic Entrepreneurial Firms" *Journal of Business Venturing 2,* 4 (Fall 1987): 285–97.

Laurie Garrett, "There Are Problems," *SRA Journal* 17, 2 (Fall 1985).

Denis Gray, Trudy Solomon, and William Hetzner, *Technological Innovation Strategies for a New Partnership* (Amsterdam, The Netherlands: Elsevier Science Publishers, 1986).

K. W. Heathington, Betty S. Roberson, and Ann J. Roberson, "Commercializing Intellectual Properties at Major Research Universities: Income Distribution," *SRA Journal* (Spring 1986).

George Kozmetsky, R. Smilor, and E. Chamberlain, eds., *Economic Development Alliances: Major New Relationships for Scientific Research and Technology Commercialization* (Austria: IC² Institute, University of Texas, 1987).

Quentin W. Lindsey, "Industry/University Research Cooperation: The State Government Role," *SRA Journal* 17, 2 (Fall 1985).

William H. Mobley, "The Role of Resources, Rewards, and Rigidity in Balancing University Teaching, Research, and Entrepreneurship," *Proceedings of 2084: A Prologue, The 1984 TEES Research Conference,* Texas A&M University, College Station, January 12, 1984.

William C. Norris, "Cooperative R&D: A Regional Strategy," *Issues in Science and Technology* (Winter 1985): 92–101.

Frank Press, Presentation by the President of the National Academy of Sciences to the Select Committee on Higher Education, Senate Chamber, State of Texas, Austin, September 11, 1986.

Bernard D. Reams, Jr., *University–Industry Research Partnerships: The Major Legal Issues in Research and Development Agreements,* vol. 1 (Westport, Conn.: Greenwood Press, Quorum Books, 1986).

Report of the White House Science Council, Panel on the Health of U.S. Colleges and Universities, February 1986.

Raymond Smilor, G. Kozmetsky, and D. Gibson, eds., *Creating the Technopolis: Linking Technology Commercialization and Economic Development* (Cambridge, Mass.: Ballinger Publishing, 1988).

9

Ramifications of Operating a Business and Industry Development Center as an Auxiliary Enterprise

Henry C. Kowalski

GMI Engineering and Management Institute (formerly General Motors Institute) in Flint, Michigan, initiated the Business and Industry Development (BID) Center in July 1983. With initial financial support from the Mott Foundation, a survey of businesses in Flint/Genesee County was undertaken to provide Mott and GMI with an assessment of the problems facing local industry. The survey indicated a specific need for various types of technical services. Completed in December 1983, it conclusively showed the need for a BID-type agency designed to spur economic development and diversification. Results of the survey were utilized to develop a three-year plan for providing the needed technological services.

The Mott Foundation—responding to a GMI proposal—pledged financial support in excess of $1 million over a five-year period in partial support of the BID Center, which would act as a readily accessible resource of engineering, manufacturing, and other business-related services to prospective and developing enterprises.

The BID Center utilizes the facilities of GMI Engineering and Management Institute to assist fledgling entrepreneurs and evolving firms in their development, adaptation, or expansion. BID is a means of following an

165

idea, product, or service from concept through all of the necessary development stages, and applying valuable assistance along the way. Its primary purpose is to facilitate economic development in the Flint/ Genesee region. From Mott's perspective, this objective closely matches the foundation's mission statement and parochial interests. Furthermore, by focusing on an economically distressed area, the BID Center was envisioned from the start as being within the federal tax-code regulations governing a Section 501(c)3 activity.[1] Sixty-six percent of BID's clients come from the immediate Flint/Genesee area, while 75 percent are within a 35-mile radius of Flint—nicely qualifying the client base. Local major outlying cities such as Lansing, Detroit, Port Huron, and Midland are more than 50 miles away. While 95 percent of BID Center clients are from within the lower Michigan area, the center has received inquiries and served clients from other regions in Michigan and other parts of the United States—even as far away as Dallas, Texas.

Background

GMI Engineering and Management Institute—a pioneer in cooperative education—is an accredited college offering bachelor of science degrees in electrical, industrial, and mechanical engineering, and management systems. Under the cooperative system of education, students alternate between 12-week periods of study on campus and a related 12-week work experience with one of the more than 250 companies participating in the program.

The institute's nearly 3,000 students come from more than 1,100 secondary schools in the United States, Canada, and five other countries. Since GMI's inception in 1924 as a degree granting four-year cooperative program in engineering and management, the vast majority—slightly more than 80 percent—of its 20,000-plus graduates have chosen to continue pursuing careers in the industrial sector, notably in such related functions as engineering, manufacturing, and management.

General Motors Corporation agreed to underwrite the school in 1926 and extended its concept to all units within the corporation. In 1945 the requirement of an undergraduate baccalaureate fifth-year thesis was initiated, thereby anchoring GMI's commitment to cooperative education and cementing its close relationship with industry.

In 1982 GMI's unique relationship of 56 years changed from its being a wholly owned subsidiary of General Motors to a privately endowed college of engineering and management. Its ties with General Motors remain strong. More importantly, however, GMI's involvement with the industrial sector in general—but especially manufacturing—is even stronger.

The gradual change from a wholly owned subsidiary to an independent institution took place over a three-year transition period. During this time, GMI began to undertake new programs of interest for its constituency, as well as fuse relationships with other industrial entities—including government. For example, in the fall of 1982, GMI began a media-based graduate program leading to a master of science degree in manufacturing management.

Elements of the Bid Center Concept

GMI's three main buildings—Academic, Campus Center, and Residence Hall—are located on a modest 45-acre campus. The Academic Building, which is more than 400,000 square feet of classrooms and halls, constitutes the undergraduate instructional facility. Approximately a quarter of this area is devoted to laboratory space—highly sophisticated and exceptionally equipped facilities—endowed through General Motors. For an undergraduate institution with nearly 3,000 students of which less than half are on campus at any one time, this means that every student can have almost 100 square feet of his or her own prime laboratory space—a luxury that is the hallmark of GMI's hands-on approach.

Based on time available during normal business hours and GMI's 48-week academic year, it has been estimated that GMI's laboratory facilities have a utilization rate of 40–50 percent (exclusive of certain computer facilities that have nearly a 100-percent utilization rate). If the use of GMI's facilities are extended over a 16- to 24-hour period, then the utilization rates of specialized laboratories approach zero. Thus, many of GMI's highly specialized facilities are also highly underutilized.

We may say, then, that the GMI circumstances contained the following resources:

1. A vast pool of creative professionals within the community;

2. An underutilized technical facility with excellent equipment;

3. A staff of industrially oriented faculty and technicians;

4. A close and unique association with industry;

5. Extensive expertise in industrially applied research and development;

6. A new independence and pursuance of recognition; and

7. A renewed commitment to the community.

Acting on its circumstances—therefore—and with support from the Mott Foundation, GMI created the BID Center to provide access for entrepreneurs, start-up companies, and developing companies to GMI's industrially oriented faculty, staff, and extensive laboratory facilities on a noncompetitive basis. A cost-effective means was thus set up to perform prototype or product evaluation, design, analysis, testing, and fabrication, as well as other business activities such as market analysis, cash flow analysis, and business plan preparation.

A typical inquirer at the BID Center is given a questionnaire (see the end of this chapter), which solicits specific information concerning the potential client's product or service and also spells out the extent of GMI's obligation and involvement. Responses to the questionnaire are utilized to ascertain GMI's position in assisting the potential client. When appropriate, an interview is then conducted between the inquirer and key faculty and staff. Should GMI decide to pursue the project, resources and time schedules are established. Subsequent action is expressly tailored to each project. The general emphasis of the BID Center is to assist—not to undertake work for the client. A marriage of faculty, staff, and client is necessary to the successful completion of a project.

Attorneys (especially patent lawyers), accountants, and other consultants may be brought into the process. Although GMI itself dispenses no seed, growth, or venture capital, the BID Center is in a position to recommend funding sources and to present potential investors with possible deals for consideration. Fees for BID Center services are arranged on a case-by-case basis. The Mott funding is used to cover GMI's out-of-pocket expenses; however, most clients do not have sufficient financial resources to pay for all of the services needed to turn their idea or product into a commercial reality. Various payment options are available, however—once a formal agreement has been initiated—which might include bartering, giving GMI a royalty or an equity position, a deferred payment plan, any combination of the above, or some other mutually agreeable arrangement. BID's goal in payment negotiations is always to provide GMI with a potential downstream income from the success of newly seeded start-up or developing companies.

Operational Procedures

Organized to be part of an economic development system in Flint/ Genesee County, the BID Center sought the advice of other experts, academic institutions, and legal counsel on how best to operate so as to minimize any potential liability—particularly in jeopardizing its 501(c)3 status. Research indicated that most business, industrial development, and innovation centers or institutes affiliated with an academic institution emphasized the educational aspects of entrepreneurship and business principles, and/or research associated with entrepreneurship or intrapreneurship. The transfer of knowledge and the sharpening of skills is a major activity. And focusing its major effort on education or basic research qualifies the academically associated center or institute as a 501(c)3 organization.[1]

The traditional university mission in research is to add to a field of knowledge. Basic research is undertaken primarily to promote an atmosphere of inquiry and to stimulate intellectual curiosity, thus expanding the frontiers of knowledge and human endeavor. U.S. academic institutions have become expert at promoting, expanding, and undertaking basic research.

GMI had carefully considered all these factors when it decided to proceed with the BID Center. Nevertheless, the operational procedures GMI chose to implement deviated from those of the academically affiliated centers and institutes consulted. In fact, its objective to establish and operate the BID Center as a provider of technical and manufacturing expertise as well as other business services to entrepreneurs and developing enterprises put it and the center clearly in a vulnerable position of nonconformance relative to Section 501(c)3 of the tax code.

GMI's overall strategy was never to be in a position where it might be viewed as part of the management in any spin-off or downstream-income arrangement with a client. (In fact, the only equity positions it has ever acquired were for less than 10-percent ownership.) Furthermore, realistic cash flow analyses showed that the BID Center would not begin to generate any substantial income for at least seven to ten years after its initiation. Actually, a ten-year period of decreasing external funding was anticipated, before self-sufficiency could be achieved.

Established companies able to underwrite an applied research-and-development program would not be processed through the BID Center. These companies are directed through GMI's Office of Research and Development and Graduate Programs. Consequently, GMI felt it could rightly claim that the BID Center serves only indigent industrial clients.

A total systems approach to economic development was initiated in the community, with the BID Center an integral part of this economic system. Other elements included an incubator and an outlet for capital funding (either through loan-based programs or venture capital funds). A flowchart depicting an idealized version of BID Center operations is contained in Figure 9.1. Its organizational structure is given in Figure 9.2. The organizational system was constructed to take advantage of influential, successful, and knowledgeable people from the community who could advise the BID Center on policy. An executive board consisting of GMI personnel, faculty, and administrators was also assembled to assist in managing the appropriate distribution of GMI resources among projects determined to be commercially viable and potentially successful.

Should the BID Center concept prove to be financially successful, the strategy was to incorporate it as a for-profit auxiliary business at that point in time, since GMI would then be receiving income from unrelated business activities that would not qualify under Section 501(c)3.

The key point is that GMI was actually taking on related business activities and research-and-development functions of an applied nature in order to assist specific entrepreneurs and specific developing companies. Furthermore, in exchange for its contribution of resources, staff, and facilities, GMI was receiving a vested interest in some of the spin-off entities it was helping to create. On the altruistic side, GMI hoped to assist in truly diversifying and developing the economy of Flint/Genesee County, which had been devastated economically by the loss of blue- and white-collar jobs in the automotive industry—Flint/Genesee being umbilically attached to General Motors.

Results

Between its inception in July 1983 and the end of December 1986—the formal funding period—the BID Center received more than 500 inquiries. From these contacts, 401 potential clients were identified, resulting in 97 projects that required some form of technical or related business assistance. Extensive, applied industrial research and development was undertaken for 43 of these projects, and led to the creation of 13 new business start-ups. In addition, 19 small to medium-size ongoing or developing businesses were assisted with technical expertise. Approximately 175 of the 401 potential clients had ideas, products, or services that may be classified as technology driven—which was the focus of the BID Center plan. ("Technology" may be defined here as the application or reduction

Figure 9.1   Flowchart of the Idealized Operating System of the BID Center in Flint, Michigan

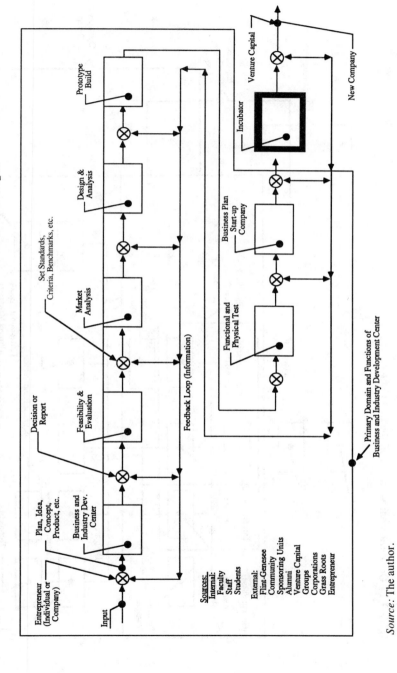

Source: The author.

Figure 9.2   Organizational Structure of the BID Center

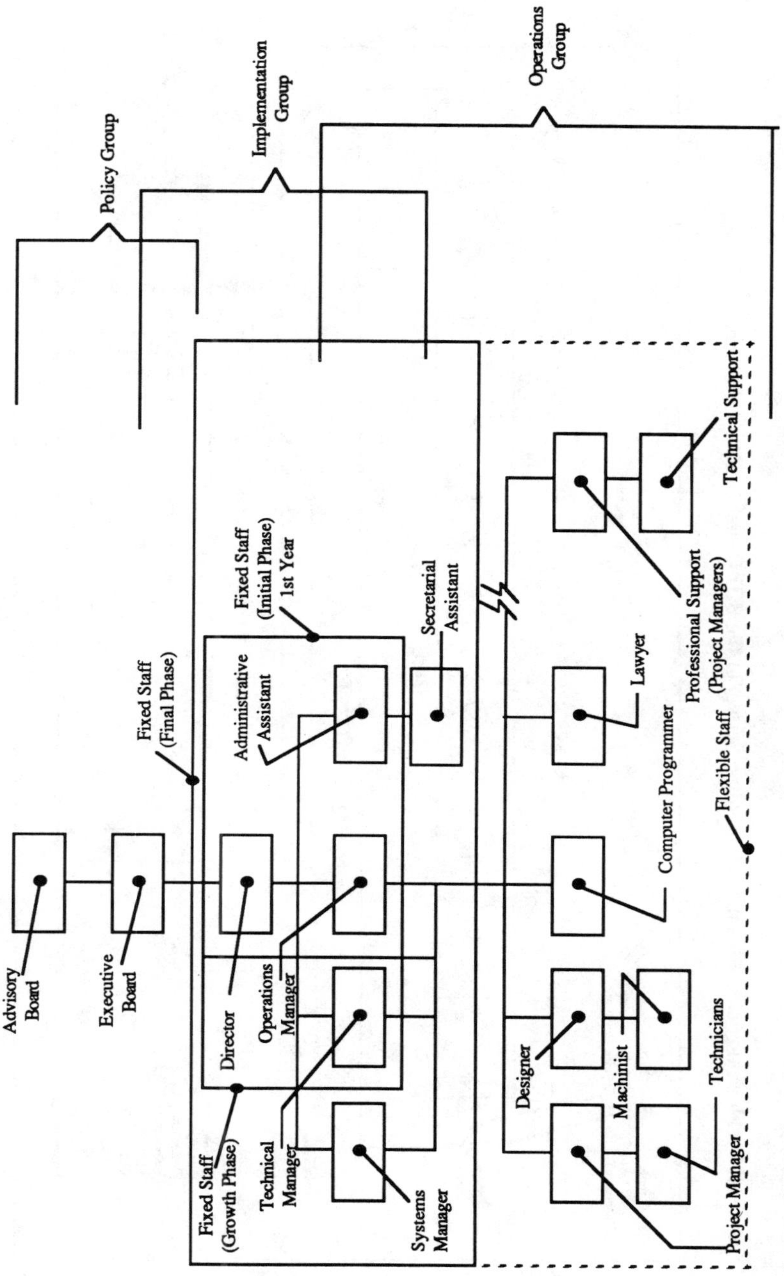

Source: The author.

to practice of techniques founded on scientific principles, practices, or concepts.) For the size of GMI and the community, this is an extremely high level of interest.

Viable ideas or proposals that cannot be served by GMI are referred to other resources in the community—which includes a community college, a private business and technical college, a small business center, an economic development corporation, and a chamber of commerce. The community college, small business center, and chamber of commerce are very active in providing a myriad of educational courses for entrepreneurs and small businesses.

The surprising number of technology-driven clients substantiates the preliminary survey's indications. This is understandable, given Flint/Genesee County's heritage as a General Motors manufacturing community—with an existing pool of highly skilled middle-management people. The design, engineering, and research departments for Buick and AC Spark Plug are also located in Flint. When the automotive industry shifted direction, many of these highly qualified professionals took advantages of early retirement along with its so-called Golden Handshake incentives. Put simply, then, these upper middle-age professionals were becoming bored. They had begun to exercise their creativity by developing innovative products or services.

Of the 13 new business start-ups that resulted from BID Center activity before 1987, nine are still in existence, while four have changed ownership. Two of the 13 companies have grown faster than their business plan projected; one is now sponsoring two students in GMI's co-op program.

Another direct effect of the BID Center has been the initiation of advanced industrial research and development on a contractual basis with five different companies: Two are in a start-up mode, while three are medium-size companies. The direct outcome of these development contracts for GMI is an infusion of $250,000 per year of income. The technologies involved in the work include gait analysis, noninvasive mass-flow metering, multiratio power transmission, physical exercising, and real-time display of video and sensor data acquisition. These projects resulted in two patents being issued to GMI faculty and staff, five patents in the pending stage, and seven more in various stages of preparation for prosecution. From this collection of newly created intellectual property, two of the patents pending have already been licensed to existing and developing companies.

Other major initiatives included a partnership with the University of Michigan–Flint on two projects. One of these resulted in the procurement of the only SBIR grant for a start-up company in the Flint/Genesee area.

In a joint venture connecting Saginaw Valley State University, Northwood Institute, and GMI, the Eastern Central Michigan Inventors Council was formed—an informative self-help network of inventors and entrepreneurs. The formation of a close relationship with three venture capital groups from the mid-Michigan area was also a major accomplishment. Other activities included seminars on venture capital and on writing business plans, establishment of an internship program for Dutch engineering students in their last year of studies, guest speaker programs, and a symposium on product liability insurance.

The bottom line for the first three and a half years of the BID Center is that its growth and success far exceeded all expectations. A steep learning curve was achieved, and the usual pains associated with uncertainty and growth gave it a dynamic and extremely visible role throughout the entire community.

Operational Problems

In April 1986 the BID Center's operation was reviewed as part of an extensive IRS audit of GMI. As a newly formed independent institution, the BID Center had not only anticipated the audit, but welcomed it. A very extensive review was made of GMI and Mott files pertaining to the BID Center grants awarded in 1983, 1984, and 1985. Following this process of discovery, the IRS agents concluded that GMI's tax liability from the BID Center operation was nonexistent during that time, essentially confirming the strategy taken by GMI in setting up the BID Center. However, the agents did rule that the BID Center was organized primarily to promote business, rather than exclusively to accomplish the charitable objectives listed in Section 501(c)3 of the tax code.

Alleviation of poverty—as contrasted to improvement of the economy in a devastated area—has to be the stated and proven interest of an institution in order for its operations to qualify as 501(c)3 activity. According to the IRS agents, the arguments that the operation of a nonviable business by a charity should be considered charitable is at odds with the developed body of law concerning charity, since the operation by any charity of a business showing a loss would convert an ordinary business activity into charity. Therefore, the government's position was that Mott's grant to GMI's BID Center did not advance purposes within the meaning of Section 501(c)3.

The IRS concluded that Mott's grant in support of BID constituted a taxable expenditure. Consequently, Mott was liable for tax of 100 percent as well as a 10-percent penalty. In remedial action taken to resolve the tax levy, GMI returned $700,000 to Mott. The IRS penalty—assessed at $70,000—was not waived. Reserving its position at the time of the IRS ruling, the Mott Foundation has subsequently gone on record as objecting to the ruling. Neither GMI nor Mott admit to any improprieties, nor do they agree with the IRS conclusions.

To resolve the problem, the parties on the defense have chosen a two-step approach. First, Mott paid the penalty under protest and then proceeded to file suit against the IRS in federal court. A long but potentially precedent-setting legal action is now in progress. Second, through the efforts of GMI and congressional representation, legislative action has been recommended that would absolve Mott from any tax liability for the grants given to the BID Center in 1983, 1984, and 1985.

Current Operation and Prognosis

Mott's support for the remaining two years of the proposed period of support had to be terminated, placing GMI's BID Center in a very precarious position. The alternatives were either to terminate the BID Center or else to continue its efforts on a drastically reduced scale, using GMI's internal resources where possible but relying primarily on income generated for services. In view of BID's success in generating applied research-and-development contracts for start-up and developing companies wherein the start-up or company was seeded by financial investors, the second alternative was considered feasible as well as most desirable. BID has been able to withstand the drastic reduction in funding and has even begun to grow slowly. Mott and community leaders continue to see BID's activity as a viable avenue to economic diversification and development.

A plan is now being developed that will allow GMI to spin off two subsidiary organizations. One is intended to be a for-profit entity that would provide more flexible access to GMI's resources of faculty, staff, and facilities. The second entity will be a charitable institution with the primary function of conducting basic research. Funding for the not-for-profit (charitable) organization would come from grants and basic research. The for-profit entity—incorporating BID—would be capitalized from the sale of equity. Business leaders in the community have already expressed an interest in this organization. The logical next step in continuing the BID Center activity is as an auxiliary enterprise at arm's length from GMI.

Conclusion and Recommendations

GMI's experience shows that, under the proper set of conditions, the resources of an academic institution can promote economic development within its constituency and community on a quid-pro-quo basis. No general conclusion can be drawn, however, pending the duplication of those unique conditions at other academic institutions.

There are a number of other observations and conclusions that can be made, though. First, a BID Center activity offers an alternative to those creative, innovative, and dynamic faculty members who have much to contribute but are not necessarily suited or comfortable with the publish-or-perish syndrome of basic research that underpins faculty promotion and ranking at most traditional academic institutions. Second, this type of activity should be part of an entire economic development system consisting of: (1) extensive business resources; (2) an incubation facility; and (3) capital.

Extensive business resources are needed for marketing analysis, business plan assistance, and operational procedures—not the least of which involve legal matters. An incubator area in which a true start-up company can have access to affordable support services is very important. Capital is extremely vital. However, to successfully conclude a deal involving seed or start-up capital, a venture capital mentality must first be cultivated within the community. Both the entrepreneur and the investor need to understand the process. The BID Center—although quite successful—was hampered in its operations because of inadequate incubation facilities and a lack of venture capital deals.

A liberal policy of faculty reward for participation in a BID Center–type of activity—especially in terms of rank—is also paramount. The development and deployment of technology in an industrial and commercial environment is by no means a mundane or trivial matter. A considerable amount of skill and discipline is necessary to produce a commercially successful venture.

If industry, academia, and government are serious about economic development and diversification in the United States, then appropriate legislation must be enacted to expedite the process. Such legislation should essentially exclude an academic institution from any tax liability on income derived from the utilization of its resources—including staff—in assisting spin-off or start-up companies to develop within its constituency and community. Interactions between the academic institution and developing enterprises necessitate making business arrangements. However, the institution is in essence investing in its own future, in the future of its constituency, and in the future of its community. The ultimate outcome turns out to be a win-win situation for all parties concerned. The income may be used to further endow the institution or to assist in covering the cost of its ongoing expenses.

The freedom of the traditional university to originate, interpret, and disseminate knowledge is a critical custodianship that it considers sacrosanct and will safeguard against all outside pressures it considers to be improper. Within the realm of academic freedom dwell not only the cherished right to study and teach subjects free of outside influences and directives, but also the right to pursue basic research as a means of stimulating intellectual curiosity and extending the frontiers of knowledge. The argument is often made that, as this highly developed process has served us in the past, so too it will give rise to tomorrow's technology.

Difficult as it may be to argue against the champions of academic freedom and their altruistic arguments, it is perhaps best to point out simply that the vast majority of U.S. universities exist by virtue of the Land Grant Act signed into law by President Lincoln in 1862. The Land Grant Act—better known as the Morrill Act—was passed because of the antiaristocratic climate that existed within the legislative body at that time. Congress was conditioned by a growth of reflective democracy, the influence of the middle class (especially those in industry and commerce), and a keen awareness of the importance of science and technology in developing the nation's cities and frontiers. All of these forces led Congress to make a significant protest against the system of higher education inherited from England. In that country, the university system catered almost exclusively to well-to-do young men in the careers of ministry, law, teaching, and the civil services. Only people with substantial means could afford to send their sons to these elite schools.

The Morrill Act of 1862 "created a system of financing which endowed at least one college in every state where the object of education would be . . . to teach such branches of learning as are related to agriculture and the mechanical arts." Its primary intent was to promote the liberal and practical education of the industrial classes in the pursuit of professions. The Land Grant Act did very well in advancing agriculture: We are now the world's largest producer of farm goods, having made tremendous progress in the science of agriculture. However—relative to agriculture— the "mechanical arts" have not fared so well.

Although universities have created and perfected basic research—and it can be argued that this research certainly has advanced the mechanical arts—such as academic endeavor is not ideally suited to the middle industrial class, or commerce. In fact, basic research is an organization of the elite for elite special-interest groups—not the least of which is the nation's military system. Industrial giants also benefit greatly from basic research. They, along with the U.S. government, are the only customers capable of financing this extremely sophisticated and highly technical business. Yet it is well documented that the real growth of the U.S. economy—creation of jobs, and increase in GNP—comes from small to medium-size companies. In 1987, the Fortune 500 companies had a net deficit in employment, while nearly 75 percent of the increase in our gross national product has repeatedly come from businesses with less than 150 people.

With the very survival of the United States as a world-economy player in the balance, we must allow developing companies an opportunity to interact with the resources within our higher educational system. The legislative mechanism proposed above—freeing the academic institution from tax liability on income from its research and resources—could accomplish such an objective. Furthermore, it would cost absolutely nothing in terms of taxpayer dollars. In the long run, it would even save dollars if the rewards reaped by progressive academic institutions from this incentive were used to offset the spiraling cost of education and research.

BID Center Questionnaire

The information requested in this questionnaire is necessary for consideration in setting up a BID Center interview. Please complete the questionnaire as accurately as possible. This will assist the BID Center in processing your potential project in a timely manner. No minimum level of development is required by the BID Center to evaluate your innovation or idea—primary consideration is given to technical soundness and commercial viability.

As a potential BID Center client, please note that:

1. The BID Center will exercise best efforts to keep disclosure confidential. Faculty, staff, students, and consultants having access to your file will have signed a nondisclosure agreement stating that they will keep your innovation or idea in confidence.
2. In consideration for this confidential evaluation, it is agreed that you will hold harmless GMI Engineering & Management Institute, its employees, students, and others assisting in the evaluation of your innovation or idea from any loss or damage arising out of this disclosure.
3. Materials submitted in connection with your innovation or idea may be retained by the BID Center. Therefore, it is your responsibility to

submit photocopies and duplicate material only. The BID Center will not be responsible for original material.

4. A personal interview and/or assistance beyond this initial evaluation is provided at the option of the BID Center and will depend upon the merit of your innovation or idea and the availability of GMI personnel and resources.

5. The BID Center assumes no obligation of any kind for the information you disclose in this questionnaire except that the BID Center will examine the information in due course and inform you of its evaluation and subsequent action. Furthermore, the BID Center is under no obligation to discuss or give any reason for its decision.

After reading the above statements and completing this questionnaire, please sign and date the following affidavit.

AFFIDAVIT

I have read the above statements and understand the conditions for processing this questionnaire. Furthermore, the information presented in the questionnaire is true to the best of my knowledge.

Signature _____ Date _____

Client(s) _____ Phone _____

Address _____

_____

Social Security No. _____

Where and at what time(s) is it best to contact you? (Please provide appropriate telephone number or address if different from above.)

Please describe your innovation idea: (Supplemental documentation or supporting material may be submitted with this questionnaire. The back page may be used for additional writing space.)

Why do you believe your innovation or idea is unique?

Why is your innovation or idea marketable?

What have you done to explore the market potential?

How did you learn about the BID Center? If you were referred, please specify by whom.

For BID Center information only. Please indicate the approximate amount of time and money invested in the development of the innovation or idea thus far:

_____ less than 40 hours   _____ less than $250
_____ 1 week to 1 month   _____ $250–$1,000
_____ 1 month to 1 year   _____ $1,000–$5,000
_____ more than a year   _____ over $5,000

Legal protection: Indicate current status.

_____ No protection
          _____ Innovation or idea cannot be protected
          _____ Legal protection or opinion for a patent attorney not sought

_____ Disclosure Document Program, letter submitted to U.S. Patent Office
          Date _____

_____ Patent application filed
          Date filed _____
          Attorney's name _____

_____ Patent granted
          U.S. and/or foreign number(s) _____

_____ Copyright
          Date issued _____
          Copyright number _____

_____ Trademark
          Date issued _____
          Trademark number _____

Your background:

   a. Work experience (general):

   b. Education:

   c. Other education, work, or experience relating to your innovation or idea:

Progress made to date:

_____ idea state
_____ model completed
_____ prototype completed
_____ refined engineering
_____ ready to market
_____ initial market stage
_____ completed marketing plan
_____ completed business plan

From what other organization, if any, have you sought assistance? (Please identify the name of the organization.)

_____ Small business center (SBC) _____
_____ Incubator facility _____
_____ Chamber of commerce _____
_____ Financial institution(s) _____
_____ Economic development organizations _____
_____ Innovation center(s) _____
_____ Other (specify) _____

Check item(s) that most nearly describes your current situation and requirements:

1. Assistance to bring innovation or idea into full commercial utilization through:
   _____ preliminary patent and/or literature search
   _____ design
   _____ model building
   _____ prototype building
   _____ engineering
   _____ testing for product liability
   _____ testing for final design
   _____ manufacturing
   _____ production processing

2. Business assistance through:
   _____ advertising
   _____ accounting
   _____ pricing
   _____ marketing research
   _____ distribution channels
   _____ other promotion (explain)

Note

Tax code law as it has developed over the years protects the "research university" from any liabilities, so long as it meets the language given in Section 501(c)3 of the federal tax code. Precedent-setting cases assist in defining the tax code. For example, applied research—although completed for a specific purpose and organization—can qualify as a 501(c)3 activity if the knowledge and expertise gained from the research can be shown to be of use in subsequent basic research undertaken for the advance of knowledge or for the public good. Similarly, certain business activities undertaken to eliminate blight or revive economically depressed areas can now qualify for tax exemption under Section 501(c)3. For a detailed discussion of unrelated business taxable income (UBTI) and the tax status of university activities, see Appendix C at the back of this book.

10

The Breeder: Forming Spin-off Corporations through University– Industry Partnerships

Frank J. Wilem, Jr.

Technology development requires the ingenuity of scientists and engineers who excel in their individual areas of research focus. However, the benefit of such development is realized only when the results are applied to the solution of real-world problems. This process is typically called "technology commercialization" and is every bit as important and challenging as the process of invention development. The skills required for successful commercialization are quite different, however, from those required in the technology development phase. An inability to recognize this can lead to the failure of spin-off corporations.

New start-up ventures represent an effective method for commercializing our federal and university lab-developed technologies. Many universities are realizing the benefits of this and wishing to promote it. Venture capitalists are recognizing the investment potential and wishing to finance solid opportunities. And the government has supported the trend through the passage of recent technology-transfer legislation. Numerous nonprofit, federal, and state organizations exist solely for the purpose of assisting small business development. Yet seldom do the individuals

183

within these organizations have a real understanding of the actual innovation and entrepreneuring process. One acknowledged essential, however, is that the entrepreneuring team be willing to make sacrifices and take risks.

There are no specific formulas for putting together a successful spin-off through the university–industry partnership. However, there are a number of common pitfalls that can guarantee failure; hence, they bear consideration. This chapter presents a private-industry/entrepreneurial viewpoint on spin-off corporations, describes some typical problems, and outlines one novel approach to spin-off formation.

Why University Spin-offs?

There are at least three reasons for supporting the establishment of university spin-off corporations. These include: (1) restoring the nation's international competitiveness; (2) promoting economic development; and (3) increasing funding for university research efforts. Typically, U.S. universities are fortunate if they can derive annual revenues from patent royalties and technology licenses that amount to even 1 or 2 percent of their current level of research funding. Furthermore, the vast majority of basic technologies are never commercialized. This represents a tremendous waste of resources—which this country cannot afford in today's highly competitive, global economy.

At one time, Europe received all the Nobel prizes; the United States simply concentrated on innovation, and became a superpower in the process. Today, we can look with pride at our many recent Nobel prizes—but it is the Japanese who are fast becoming the leaders in innovation. One clear indicator of the shift in innovation leadership is the number of U.S. patents awarded to non-U.S. entities. Of the 89,385 patents granted in 1987, 46.6 percent went to foreigners (19 percent of this total, to the Japanese alone). This represents a substantial increase from 20 years ago, when only 25 percent of all U.S. patents were awarded to foreigners.[1]

The key to our regaining the competitive edge lies in the practical application of the tremendous wealth of new technology resources. The United States must no longer tolerate the waste of its university-developed technology, nor wait for other countries to take the lead in commercializing it. Also, the nation must focus on maintaining its technology resource base. Currently in the United States, three lawyers graduate for every engineer; whereas Japan graduates one lawyer for every three engineers.

In the past, economic development was largely a matter of "smokestack chasing," in which one region attempted to lure industry from a second region—resulting in a zero-sum gain. While some communities are awakening to the fact that this is not in the nation's best interests over the long term, many still have not seen the light. The pursuit of economic development through small business development—establishing new start-up companies, or supporting the growth of existing business through the application of new technologies—is not a particularly easy proposition. However, it is one of the few ways to achieve real long-term growth. Small businesses account for 55 percent of all existing jobs and nearly all of the new jobs since 1980. Small business enterprise also accounts for almost 50 percent of the gross national product.[2]

The formation of spin-offs can increase research funding in at least two ways. First, there is the direct avenue through royalties, fees, and equity in start-ups. This may be followed by additional industry-sponsored research from the very spin-offs that the university created. Second, as the spin-offs continue to grow and prosper, they increase the tax base available for federally funded research.

Spin-offs: Myths and Pitfalls

Spin-offs offer their parent institution several commercial and technological advantages. However, it is also commonly recognized that small businesses suffer a high rate of failure—approaching 90 percent within the first two years.[3] Clearly, the actual process of establishing a new spin-off—while desirable—is a difficult and risky task.

The two primary causes of small business failure are generally acknowledged to be undercapitalization and/or poor management. However, these are really the effects of a larger problem: lack of understanding of the entrepreneuring process. Those who have never been directly involved in the establishment of a new start-up find it difficult—if not impossible—to understand what a monumental task it always is. They generally do not appreciate the importance of execution, relative to the initial concept.

Developing the product and business concept is the easy part. Executing the concept—or entrepreneuring—is the tough part. In the four sections that follow, some of the most common entrepreneurial pitfalls are discussed.

The Myth of the Million-Dollar Idea

One of the primary impediments to the formation and success of a new venture may very well be the myth of the "million-dollar idea." To say that there are no million-dollar ideas may be somewhat brash. Yet it is not far from the truth. Many ideas—if properly executed—can lead to the development of multimillion-dollar enterprises. This takes a great deal of hard work, however.

The problem comes when the inventing entity must attract investors and perhaps an entrepreneur/management team to launch the start-up company. Believing that the raw idea is worth a million dollars and that its execution will be pretty straightforward can cause the inventor to have unrealistic expectations. This results in a too lengthy process of negotiation; even if a deal is finally struck, the operating team may wind up with insufficient incentives to motivate them through the hard times.

The Better-Mousetrap Syndrome

The technology-oriented entrepreneur may lead the company to become technology (rather than market) driven. Typically this type of individual seeks to develop the absolute best "mousetrap," and constantly pursues perfection. Although admirable, this tendency can result in a disproportionate amount of the available corporate resources being expended on continued refinements and expansion of the basic product.

This is a very common failing for new technology start-ups. While not the approach the author would recommend, one school of thought says that, when a new product has reached the point where it is essentially working and can be manufactured, the scientist should be shot—thereby removing him or her from the picture so the company can make a profit!

It is generally recognized that IBM's personal computer (PC) did not incorporate the best technology available at the time of its introduction. Yet the product quickly dominated the market, and IBM eventually derived a significant share of its corporate profits for the sale of PCs. As with many other products, the giant electronics firm accomplished this through superior marketing strategy, using its considerable reputation to establish industry standards at a time when they were sorely needed. A good marketing approach and a quality product is at least as important as having the absolute best technology and design.

The Ivory-Tower Syndrome

Another problem experienced by start-ups is the "ivory-tower syndrome." This can happen when the entrepreneur adopts a self-satisfied attitude and projects an image of driving fancy cars and living the good life, coupled with his or her failure to understand that this is achievable only from the profits of the venture—not from the start-up capital. The stories abound of new ventures headquartered amid palatial surroundings, abundant staff, and copious quantities of the best and latest office equipment. The entrepreneur then begins to believe that success has already been won. That's when he or she loses the essential "burn" or "hunger."

As the amount of venture capital devoted to a new venture increases, so does the level of success that the company must achieve in order to generate even the necessary minimum rate of return for the investor. While a new venture should not be undercapitalized, overcapitalizing it can be just as dangerous. The venture must operate lean—and the entrepreneural team remain highly motivated—until the venture has achieved real success. Investors prefer entrepreneurs who frugally select very modest quarters: One observed that "the most important innovations seem to happen in buildings with leaky roofs." True entrepreneurs "don't care if they have to sit on orange crates as long as they have the tools to get the job done. Perhaps the leaky roof keeps away those who don't belong."[4]

The Difficult Problem of Innovation

The *Wall Street Journal* recently chronicled a problem that even Fortune 500 companies have with innovation, in an article entitled "Dupont's Difficulty in Selling Kevlar Show Hurdles of Innovation."[5] Over a 25-year period, Dupont spent $700 million, with an additional $200 million in operating losses, to develop and promote Kevlar. They had a product concept in search of a market. Now that the product is finally beginning to sell and sales are reaching the $300-million level, their patents are shortly due to expire. Dupont's difficulties illustrate that technological breakthrough is no guarantee of financial triumph. The university considering a spin-off formation had better appreciate the difficulty, risk, and effort associated with the innovation phase.

Three Steps to a Successful Spin-off

Having looked at the reasons for forming spin-offs and some of the associated problems, we may now examine the key steps in pursuing this course.

Step 1: Defining Spin-off Goals

The university's first step should be to define the goals of its technology commercialization. While this may sound obvious, a careful approach at this stage can turn up goals quite different from those that might otherwise have been adopted.

Typically, the university will have a single individual (or at best, a small group) acting in the role of university dealmaker. In the absence of a formal technology-commercialization agenda, the dealmaker is forced to establish his or her own criteria. Faced with a very real concern about how his or her efforts are likely to be viewed after the fact, the dealmaker will naturally take a hard-nosed position so as never to be accused of "giving away the store." While this may indeed represent university policy, it is not without a downside.

Most universities have an abundance of technologies with commercialization potential, but are very limited in the way of commercialization resources. Devoting a significant amount of these scant resources to lengthy negotiations on one or two deals can mean that other technologies remain in the lab. This leads to two problems.

The first problem stems from the fact that technology is perishable. Today's fast pace of technology development causes viable opportunities to be lost as new advances occur. This allows only a narrow window of opportunity for commercialization. The second problem is that picking the winners from a group of technologies remains a highly subjective proposition. It may very well be that some of the technologies selected for commercialization will lack the success that certain of the "losers" might have achieved, had they made it to the marketplace.

Recently an informal experiment was conducted to determine if there could be developed a model that would pick the one or two winning technologies from a large group with a reasonably high degree of confidence. In this experiment, several groups were presented with a number of potential business opportunities and asked to evaluate them. The groups included representatives from a wide spectrum including investors, venture capitalists, and CEOs.

It was expected that, while there would be great differences between individual appraisals, there would also be an overall correlation within the group—indicating one or two clear winners. The final results, however, indicated no such correlation. Selecting a few optimum commercialization candidates is therefore a tough process.

One approach that the university might consider is to strive for "maximum deal velocity." This entails trying to commercialize as much of the university's technology as possible, accepting the fact that in some instances this may mean settling for less than might otherwise be possible. In doing this, the university hopes that reducing the amount of time and effort it takes to structure each deal will result in more deals and less "spoilage" of technologies that might otherwise miss their window of opportunity. Such a strategy has several effects.

First of all, as the university's share goes down, the pieces left over for the entrepreneur and investor go up. It becomes easier to latch the deal together, and the commercializing entity has a higher incentive. More deals mean more visibility, which enhances the possibility for commercializing future technologies.

Other goals of maximum deal velocity may include ensuring that the technology becomes commercialized so that our society will receive its associated social benefits: more jobs; patriotism-supporting national competitiveness; and good public relations, all around. NASA has certainly focused on this last benefit as a technology-transfer goal. Its Technology Utilization Program is intended to demonstrate the spin-off benefits of the space program. Universities can achieve a similar PR advantage by demonstrating the spin-off benefits of federally sponsored university research.

Step 2: Recognizing the Entrepreneurial Function

Regardless of the product, the specific industry, or the sophistication of the technology, the success of a start-up venture is dependent on how successfully the role of the entrepreneur is handled. The entrepreneurial process has a unique set of problems that are really quite hard to understand and appreciate without having experienced them firsthand. This is evidenced by the fact that 90 percent of all businesses that fail do so because of the management team.[6]

An entrepreneur has to be single-minded and have a burning desire to succeed. Many—probably the majority—of viable small start-up businesses owe their success to a highly motivated and aggressive entrepreneur who is fully willing and able to meet with venture capitalists in the morning, type up his or her own correspondence in the early afternoon, and then take out the trash before going home. The point is that the entrepreneur must generally stretch what there is much further than most other people would like. He or she has to take advantage of the resources at hand and be willing to get down in the trenches and do the dirty work, as well. The work is certainly not glamorous, but it can be rewarding.

So when a university is ready to promote the establishment of a new spin-off corporation, who is there to fill the demanding role of entrepreneur? One obvious choice is the researcher who developed the technology around which the spin-off is intended to be formed. Yet this approach has some serious drawbacks.

Scientist/researchers understand the need for specialized skills in developing the technologies with which they are involved. A certain individual researcher may represent the best choice over anyone else in his or her entire field to be involved in the development of a particular technology. However, it takes an equally important though completely different set of specialized skills to be an entrepreneur and successfully convert the basic technology or product concept into a successful business.

The transition from researcher to entrepreneur requires a dramatic change in culture. Researchers seek truth and an open exchange of ideas. They must be allowed to follow their instincts, explore with the freedom to fail, and not be overly pressured to show near-term progress. Entrepreneurs—on the other hand—must seek the single solution that will maximize profitability, be concerned with near-term results, make compromises in the interest of expediency, and seek market acceptance.

There is still another problem in the researcher's assuming the entrepreneur's role. Commonly, the entrepreneur is compensated through an equity participation in the new start-up company. However, this form of incentive may not work quite so well with researchers who are ingrained with the publish-or-perish concept. They may prefer to receive professional recognition for their efforts through publishing, rather than financial compensation. Often, however, publishing may be in direct conflict with the need to retain proprietary information for the new start-up company. Thus, a researcher involved in the formation of a spin-off company must be prepared to accept the possibility of financial compensation in lieu of professional notoriety.

The researcher-entrepreneur immediately encounters numerous learning curves that he or she must surmount. Failure in even a single key area—be it marketing, manufacturing, finance, sales, or management—will generally result in failure of the entire business. The high failure rate of small businesses shows how hard it is for an entire team—much less a single individual—to meet all of these demands successfully. It is a very rare person indeed who, with no prior experience in the start-up and operation of a small business, can wear such a wide range of hats.

There may be an axiom in all this: Good researchers do not good entrepreneurs make. A better approach might be for the scientist or engineer to focus on those areas in which he or she excels, and look for someone else to fill the entrepreneur's role. Spin-offs formed through joint ventures between universities and private industry may be one alternative. The idea is to combine the skills of experienced entrepreneurs in the for-profit sector with the technology development talents of the university staff. This approach is more efficient and risk reducing, since it makes use of proven resources.

Step 3: Structuring the Deal

The long-term success of a new venture is contingent on all involved parties perceiving that they have received fair and equitable treatment in the formation of the spin-off corporation deal. First of all, the university should consider its projected ROI (return on investment) when negotiating a deal. If the research is publicly funded, the university has not—in fact—put out a lot of investment.

When structuring a university's return from the spin-off, it should be recognized that this is a long-term proposition and that some ventures— maybe even most—are going to fail. This may lead one to conclude that the university should take its cut at the front end. If the technology is being sold to a well-heeled corporation, this may be possible. However, with a start-up company, up-front fees or early-stage royalties place a heavy burden on the new venture's cash flow. They also conflict with the investor's interest in applying the venture capital toward growth of the company. Moreover, in taking its cut up front, the university is diminishing its risk and must therefore be prepared to accept a reduced share. Perhaps universities should consider a strategy similar to that of the venture capitalist, who typically looks for a return when the company goes public or when a buyout occurs.

The second thing to consider is that the entrepreneur must have substantial incentive to succeed. The university—in figuring out what it *can* get out of the deal with the entrepreneur—had better think about what it *should* get, instead. That is, in the excitement of launching the company, the entrepreneur may be willing to settle for less than what may really be fair. However, later—when the going gets tough—he or she may lack the necessary motivation to continue the struggle. A win-win situation should always be the goal.

The Breeder: A New Approach to Spawning Spin-off Companies

Investment bankers understand finances and are well equipped to interpret complex financial statements. University professors achieve recognized status as experts in their areas of specialization, and know how to compete for research funds. For their part, state organizations attempt to implement programs that are intended to meet broad programmatic objectives for small business development, but they seldom meet the individual needs of the entrepreneur. The Gulf Coast Breeder in Pass Christian, Mississippi, was established to serve the needs of the entrepreneur and act as a bridge connecting the parties vital to the technology commercialization process.

The "breeder" is a novel approach that focuses on the systematic development of innovation and entrepreneuring. Since its founding in 1987, the Gulf Coast Breeder's mission has been to provide the business, entrepreneurial, and government interfacing required for successful technology transfer. It was built to handle everything from management, marketing, and manufacturing, to securing venture capital resources.

Breeder Background

The breeder concept was developed by a group of successful entrepreneurs who were experienced in establishing and operating profitable technology-oriented companies. The Gulf Coast Breeder team has a strong technical background as well as a firsthand familiarity with the difficulties of commercialization. The Breeder is based on the assumption that a structured approach to the commercialization of technology should be far more successful than the typical random approach. It was this assumption—coupled with the recent evolution of new concepts for developing start-up businesses, such as incubators—that led to implementation of the breeder concept.

Nonprofit organizations have a very questionable record of success when it comes to developing new for-profit ventures. There are very few exceptions to this. The problem is that the nonprofit organization itself lacks the very incentives necessary to form new start-up corporations. If an organization is not for profit, then the question must be asked: What is it for? The strength of the breeder concept is that it is structured around the same incentives as the start-up.

Breeder Purpose and Concept

The primary purpose of the Gulf Coast Breeder is to support the development of technology based start-up companies. Product concepts can come from several sources, including university or federal lab-developed technologies. Its goal is to develop spin-off companies that address highly profitable niche markets and to establish these companies quickly, efficiently, and with an absolute minimum amount of risk.

The Breeder is founded on two simple principles: First, any organization whose focus is to support new venture creation and development must be managed by individuals who have a complete firsthand understanding of the innovation and entrepreneuring process. Simply stated, they must have been personally involved in a new start-up. Second, if it is to make a meaningful impact and survive the long term, such an organization must be a for-profit entity.

The Breeder is operated by a team of experienced entrepreneurs with a proven management, marketing, financial, manufacturing, and technical track record and with experience in successful technology transfer. The Breeder focuses on innovation—not R&D—pursuing the commercialization of existing technologies about which there are few, if any, open questions of feasibility.

The Breeder is privately funded. It has the ability to provide whatever pieces are missing for the commercialization of a product concept. These pieces may be in the areas of marketing, management, finances, or manufacture. The Breeder treats each technology or idea as unique, and uses a flexible commercialization approach that can match the "style" of the particular technology.

The Breeder Process

Product concepts are evaluated in a four-step process. In the first step, a cursory analysis is performed to identify those concepts that seem most promising and best suited for the Breeder's resources. After this Phase I evaluation, selected candidates enter Phase II in which any open questions such as market size, technical feasibility, and manufacturability are addressed. At the conclusion of Phase II, the product concepts are once again evaluated—this time in light of the Phase II results. If the product still looks viable, it then enters Phase III in which a business plan is developed and financing obtained. In Phase IV, the product concept is taken to market.

Through the Breeder process, product-concept commercialization may be pursued in a number of different ways. One approach is the formation of a freestanding start-up company that would operate in a completely autonomous fashion. Another approach involves the Breeder providing some operational resources directly or through one of its partner companies, thereby reducing start-up cost and risk. For example, existing marketing channels, manufacturing capabilities, or existing management talent may be applied in order to facilitate the Breeder process. In a third scenario, the product concept may be commercialized exclusively by one of the Breeder partners or a Breeder affiliate. Or the Breeder may function in a brokering capacity, helping to transfer the technology to an independent private company.

The Breeder derives its revenue through up-front licensing fees to the start-up companies, royalty payments, equity positions, or a combination of all three.

Breeder Association

Many companies are concerned with maintaining a competitive edge, but have limited resources. Since only a small part of the technology that surfaces as a result of the Breeder's efforts can be commercialized by its partners, a Breeder Association has been formed. There is no membership fee, and the association includes a number of companies that have an interest in applying technology either to augment their product lines or to improve their manufacturing processes.

In order to be successful, it is imperative that the Breeder's efforts be focused on a few discrete areas of technology. However, technology that is not of interest to the Breeder may be of interest to association members. When this occurs, the interface is handled on a brokering basis. This approach leverages the technology-transfer efforts of both the Breeder and the United States in general. In the Breeder's case, it provides an additional source of revenue to help offset operating costs. In the case of the nation, the Breeder's interaction with a large number of small businesses greatly improves the prospects for widespread utilization of existing technologies.

Notes

1. *Insight,* April 4, 1988, p. 26.
2. *Entrepreneur Magazine* (April 1988): 13.
3. *Hattiesburg American Newspaper,* April 2, 1988, p. 11A.
4. Gifford Pinchot III, *Intrapreneuring* (New York: Harper and Row, 1985), p. 73.
5. *Wall Street Journal,* September 29, 1987, p. 1.
6. George Kozmetsky, Michael Gill, and Raymond Smilor, *Financing and Managing Fast-growth Companies* (Lexington, Mass.: Lexington Books, 1986).

Part IV

Turning University Research into Business Opportunities

11

Technology Commercialization in Illinois

Demetria Giannisis, Raymond A. Willis, and Nicholas B. Maher

The status of the United States as technological innovator and producer had been called into question even before it became a pivotal issue in the 1988 presidential campaign. The 1985 Presidential Commission on Industrial Competitiveness stressed the role of technological innovation, productivity growth, and human capital as structural indicators of competitiveness. The report's analysis progressed from the immediate context of "commercializing new technologies through improved manufacturing" to the extended context of reducing the federal deficit, improving school curricula, and revising the tax system to encourage innovation. The commission's strategy included the following recommendations: Create a "solid foundation of science and technology relevant to commercial uses"; apply "advances in knowledge to commercial products and processes"; and "protect intellectual property." These basic challenges and goals continue to define the agenda of those working to promote technology transfer and commercialization.

Federal Laws Supporting Technology Transfer

Prior to the president's commission report, federal laws directly impacting technology transfer had already been created or refined. The Patent and Trademark Amendment Act of 1980 included substantive changes in federal patent law and allowed universities, small business, and not-for-profit institutions the option of retaining title to inventions developed with government funds. The 1980 passage of the Stevenson–Wydler Innovation Act was specifically designed to promote cooperative research and granted federal laboratories the authority to license and publicize commercially relevant inventions.

The Economic Recovery Tax Act of 1981 was passed as part of the Reagan administration's tax legislation program. This act included a range of provisions designed to stimulate industrial interactions with universities, including a 25-percent tax credit allowance for companies with increases in research-and-development expenses above existing levels. The bill also allowed credits to companies that invested in equipment for research-and-development purposes. The original bill expired in 1985, but then was extended through 1986 with a slight reduction in the tax credit down to 20 percent. Presently, industry–university associations such as CORETECH are lobbying to make the act a permanent law. Additional financial incentives to enhance collaboration between previously competitive industrial firms were included in the 1984 National Cooperative Research Act, which relaxed antitrust regulations.

The aforementioned federal laws were passed to spur the development of technology-intensive industries and strengthen the foundation of research partnerships between business and institutions of higher education. In conjunction with these federal mandates, and in reaction to the impact of the mid-1980s recession, state governments pursued independent economic development strategies to exploit regional assets. A cursory analysis of state programs reveals a common agenda among them to promote small business growth and attract high-tech related manufacturing and/or productive services.

In support of these state-level actions, a recent report analyzed the status of the Midwest region's economy. The report outlines steps that state governments have followed or can pursue to revitalize their regional economy:

1. Apply new technologies to renew existing industries through specialized innovation centers such as those in Indiana and Michigan;

2. Promote new enterprise development through special financing as well as regulatory, management, and technical assistance programs like those in Michigan, Iowa, Wisconsin, Pennsylvania, and Indiana;

3. Invest in human capital through educational reforms by helping local community colleges and adult-education districts establish joint retraining and upgrading programs with the private sector;

4. Renew state infrastructure in areas where industrial productivity demands new roads, bridges, and sewers; and

5. Promote high-technology development through specialized university–industry institutes and other efforts that build on the unique strengths of the state (AmeriTrust Corporation 1986).

Leveraging Illinois' Regional Assets through the I-Tec Program

As a fair composite of the nation's economic diversity, Illinois provides a particularly relevant example of state government response to the changing national economic environment. Remarkable similarities exist between Illinois and the national occupational distribution in key areas such as manufacturing, trade, finance, transportation, mining, agriculture, and government. Parallel distributions are also found in the state's social characteristics, including levels of education, population concentration, race, and the mix of wage-earning versus salaried workers. Perhaps the most telling similarity is in terms of jobs lost: "In the past decade 200,000 industrial jobs in the state were lost to automation, labor strife and foreign competition"—a pattern mirroring the national reality (Davis 1988).

Illinois' response to the "call to competitiveness" preceded the presidential commission's report with a variety of civic actions, including the formation of a Task Force on High Technology Development in 1981 and the creation of the statewide Illinois Technology Transfer and Commercialization Center Program (I-TEC) in 1985. Leaders of Illinois' academic, business, and industrial communities formed a task force and developed a strategic plan to retain and attract high-tech industries, nurture emerging growth industries, and publicize the state's research resources. The Governor's Commission on Science and Technology was later established with public and private-sector representatives to further the task force's goals and implement its recommendations.

As a part of the state's overall plan for high-technology development, the I-TEC received roughly $1.8 million in 1985 (increased to $6 million in 1986) to support the Business Innovation Fund (BIF) program and the operation of 13 Technology Commercialization Centers (TCCs).

The Business Innovation Fund—administered through the Illinois Department of Commerce and Community Affairs' Small Business Assistance Bureau—offers royalty financing for the development of innovative products or services in existing companies and the expansion of small (one to ten employees) technology-based businesses. To be eligible for a BIF award, the proposed expansion or product/service development activities must be conducted in cooperation with an Illinois university, college, or not-for-profit research organization, must have its principal place of operation in Illinois, and must be at least 51 percent owned by an Illinois resident. The fund sponsors no more than 50 percent of the planned costs, and contributes up to $100,000 per company.

Technology Commercialization Centers were established at 13 major research institutions across the state. The TCC program goal is to capture and catalyze the transfer of innovation research products from universities and federal laboratories to business and industry in Illinois. Each TCC provides access to the specialized resources at its research institution and offers a variety of services relevant to the process of technology commercialization or the broader operating goal to facilitate knowledge sharing between the institution and the Illinois business community.

Although there is some variance among the TCCs, the basic services provided to clients either directly by TCC staff or indirectly through the resources that the TCC has access to includes two general categories: market assessment, and product development services. Specific services include: initial screening and assessment for marketability and patentability; feasibility testing, including technical, legal, and safety analysis; initial market analysis; funding for product development; technical assistance including prototype testing, design, and production analysis; strategic market analysis; full-scale product research and development; general legal and accounting advice; business plan assistance; incubator services; and production and distribution consulting.

To fund product development research, TCC seed funds are leveraged with a variety of other public-sector monies—including the Illinois Business Innovation Fund, Venture Fund, Equity Investment Fund, and the federal Small Business Innovation Research (SBIR) program. Projects with commercial merit receive TCC funding assistance in the form of access to university resources (labs, faculty consulting, and specialized equipment) and through grants of up to $25,000 per client for the support of research-and-development work directly related to commercialization.

The financing needs of start-ups vary with the company's development stage (i.e., research and development, start-up, expansion, workout, and buyout), so flexible funding options must be pursued over time. To meet the range of client needs, the TCC may also help to coordinate the client's search for venture capital and informal investment-risk capital. This service is particularly critical in cases where the client is searching for informal investment capital. In combination with state funding programs created to fill the gap[1] between equity financing and sources for funds under $500,000, informal investors provide a needed funding bridge for start-ups.

A recent study on the networks and investment criteria used by East Coast investors corroborates the assumption that most informal investors use close contacts (i.e., friends, other investors, and business associates) to identify investment opportunities. The study surveyed some 130 informal investors who reported that they had raised $38 million to finance 286 new venture proposals in a three-year period, either as individuals or through their networks. Both venture capitalists and private investors consistently rank the new venture's managerial capability and the product's market potential as the most important determinants of investment outcome. Other important distinctions exist, however, between the typical venture capitalists and informal investor's criteria. For example:

> Angels [i.e., informal investors] are not interested in a thorough business plan, a *sine qua non* for venture capitalists. . . . Unlike the capital firms, angels are not interested in competitive insulation. They do not limit their investments to industries that are appealing or with which they are familiar, nor do they care very much about the degree to which the entrepreneur has identified competition. (Haar, Starr, and Macmillan 1988)

In contrast to other regions, the majority of new Midwest ventures are created by first-time entrepreneurs who do not have a history of extensive management experience. As a result, they face greater obstacles in their attempt to secure financing that does not demand a large equity sacrifice. To reduce the mortality rate of start-ups and offset this trend, Small Business Development Centers (SBDCs) located throughout Illinois provide a variety of business support services, including long-term management assistance programs. These centers complement the state TCCs and the other federal Small Business Administration policy directives to foster technological innovation and commercialization, such as the SBIR grant program.

The Illinois share of SBIR awards has grown incrementally, but the state has seen a significant increase in the academic and business community's awareness of the federal grant program and the number of SBIR applications. From FY 1983 through FY 1986, Illinois small businesses received a total of 118 awards, which amounted to $10.4 million. Although Illinois received far less than the high-ranking states (California and Massachusetts), it "fared better than thirty-one states and equaled seven other states in its number of awards (Sherrod 1987).

To assess the overall impact of the TCCs, the Illinois Department of Commerce and Community Affairs has designed an evaluation program that includes measures of TCC activity and performance impact indicators. Each TCC reports to the Illinois Department of Commerce and Community Affairs on a quarterly basis using the MISTEC data base, which was specifically designed to meet the state's evaluation criteria and data-collection procedures. The TCC program's performance impact indicators include some of the following measures: patent applications applied for; patents approved; new firms started; new products or services commercialized; firm relocation to Illinois; SBIR awards applied for; and SBIR awards received.

Figure 11.1 displays the best estimates of the TCCs. As a performance measure, the indicator "jobs created and retained" assumes a direct relationship between the support of an innovative product or process and the ability of firms to remain competitive and retain (or create) jobs. It is inherently difficult to estimate the relationship between the commercialization of an innovation and the long-term direct and indirect producer–supplier linkages that might result. Employment estimates could approach higher levels of reliability if case analyses were conducted to test the assumed links and if it were then determined that: (1) a technical solution resulted from a project; (2) the solution was commercially applied; (3) the economic effects of the solution (e.g., percentage cost reduction) were of a magnitude that could be plausibly associated—in the absence of more refined endeavors such as the computation of supply elasticities—with changes in output and employment (Feller 1987).

The aggressive expansion of technology transfer in Illinois that has been achieved through the TCC program is apparent despite the fact that the results illustrated in Figure 11.2 do not include complete tabulations for 1988 and cannot illustrate the full range of TCC activities.

The University of Chicago Technology-Transfer Program

The institutional commitment of the University of Chicago and Argonne National Laboratory is demonstrated by the existence of three

Figure 11.1   **Illinois Jobs Created and Retained via TCC Activity, January 1986 to April 1988**

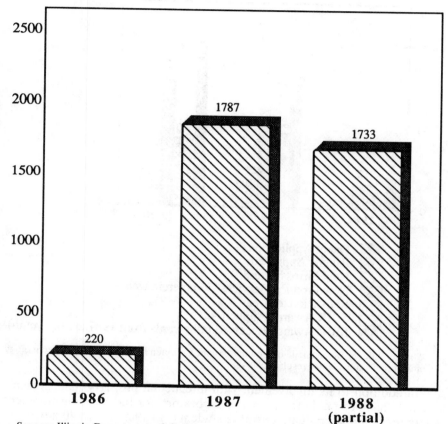

Source: Illinois Department of Commerce and Community Affairs, I-TEC Technology Commercialization Center results, 1986–88.

entities to promote the commercialization of research results generated by their combined research budgets of $389 million annually: the TCCs at Argonne and the University of Chicago, and the Argonne National Laboratory/University of Chicago Development Corporation (ARCH).

The University of Chicago's TCC was established in 1986. During that same year, the university formed the private, nonprofit, wholly owned corporation ARCH to manage the commercial development of intellectual property originating at both the university and Argonne National Laboratory.[2]

Figure 11.2   Tabulated Values of TCC Results, January 1986 to April 1988

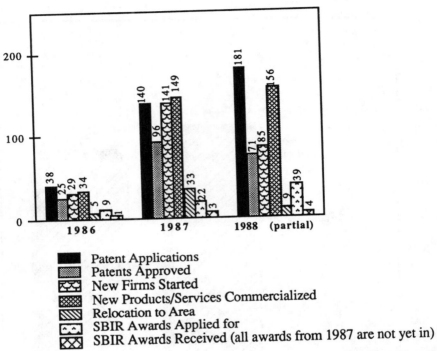

Patent Applications
Patents Approved
New Firms Started
New Products/Services Commercialized
Relocation to Area
SBIR Awards Applied for
SBIR Awards Received (all awards from 1987 are not yet in)

Source: Illinois Department of Commerce and Community Affairs, I-TEC Technology Commercialization Center results, 1986–88.

In addition to its involvement with intellectual property management, ARCH has been able to serve as a laboratory for business development activities among graduate business students. In 1988, about 40 advanced M.B.A. students with from two to five years of full-time technical, managerial, and/or scientific work experience were volunteering about 20 hours a week as ARCH associates. Under ARCH staff direction, the associates serve as market investigators for new technologies generated at either the university or the laboratory. They also provide preliminary market, patent, and competitive analysis, as well as first-draft business plans and early-stage product development strategies for spin-offs.

Three Emerging Spin-off Models

Since the establishment of ARCH and the TCCs, the number of spin-off ventures created as a direct result of University of Chicago research increased from one at that time to a total of five new firms in 1988. Every new venture created has reduced the learning costs associated with establishing a successful technology-transfer program, and increased the reciprocal quality of interaction between university faculty, administration, and the business community at large.

A spin-off may be defined as a company producing a product or service derived from research conducted at a university. Evidence from the histories of the University of Chicago's spin-off ventures suggests the emergence of three models: (1) the entrepreneurial; (2) the traditional; and (3) the institutional. Although the development path of each spin-off venture contains overlapping elements of each model, the categorization is based on the primary promoting agent in the commercialization process.

The Entrepreneurial Model

Faculty or students who have been integral to the founding of their own company may be described as entrepreneurs. In this category, the formation and growth of the newly established start-up may be largely attributed to the combination of expertise and independent motivation that the entrepreneurial faculty member has brought to the commercialization process.

The Traditional Model

The university has historically been recognized as a source of innovative ideas and technologies. In the traditional model, the commercialization potential of a university-based technology is pursued by an outside business entity. Through various avenues, that business entity approaches the inventor or institution with a proposal for development of a university-owned technology. The success of this model is contingent on the referral networks established between industry and the university, and the ease of knowledge sharing between these sectors.

The Institutional Model

As the term "institutional" implies, in this model the commercialization process is managed with the university through an organization similar to a TCC or a wholly owned not-for-profit subsidiary of the university such as ARCH. Several institutions, including the University of Utah, have used this commercialization strategy with varying degrees of success. In this category, a spin-off venture is created as a result of the university's formal process of technology identification, assessment, and development. The university works on behalf of the faculty member to reduce the amount of faculty time and effort required—and supply expertise that the faculty member may lack—to patent, license, and commercialize the technology.

The following case histories illustrate the range of possible spin-off paths.

Case Studies

Case History #1: The Entrepreneurial Model

Orchid One was founded in 1987 by Dr. Albert V. Crewe of the University of Chicago—an internationally recognized pioneer in the field of electron-beam technology, and inventor of the scanning-transmission electron microscope. The history of Orchid One attests to the fact that the innovation process is often not a neat linear progression from basic research to applied research, development, marketing, and distribution. Rather, the process often involves overlapping stages of development and reformulation.

The history of Dr. Crewe's creative contribution to electron microscopy includes work as an academic and as a consultant in the field of imaging science. Crewe saw potential in applying the technology he had used in developing the scanning electron microscope to the concept of microstorage and data retrieval. Following the normal faculty pattern in searching for funds, Crewe appealed to government agencies and other grant sources for research support. Although Crewe had successfully obtained funding for his research related to the scanning electron microscope, he was not able to secure government funding for the microstorage and data-retrieval project.

Undaunted, Crewe began to work through his own referral network. Eventually, serendipity and perseverance resulted in contact with a member of the Visiting Committee to the Division of Physical Sciences, who was interested in the project enough to invest and assist Crewe further in the search for capital. The new partners searched for out-of-state investors after encountering a consistent pattern of reluctance among local firms to fund early stage high-tech ventures. Crosspoint Venture Partners, a West Coast firm with a reputation for funding early stage start-ups, recognized the merit of the concept and proposed the formation of a limited research-and-development partnership called Electron Beam Memories to support Crewe's research program up to the proof-of-principle stage.

The concept of this limited research-and-development partnership that would include a faculty member using institutional resources was presented to the University of Chicago for consideration. As a result, in 1984 the university first entered into a contract that allowed investment capital to be received by a faculty member for research activity in much the same way as government grant funds were accepted. Under the terms of the agreement, reimbursement to the university included both direct costs and indirect costs at the approved rate for research costs under government contracts.

Crosspoint Ventures embarked on a partnership with Dr. Crewe that outlasted the original electron-beam proposal and eventually led to the formation of Orchid One. Because industry had not yet absorbed the concept of optical memory, the corporations that Crewe approached were not prepared to accept the electron-beam innovation, which demanded an entirely new orientation. Thus—regardless of the success of the first research stage—Electron Beam Memories was unable to obtain second-stage financing for development of Crewe's discovery beyond the proof-of-concept phase.

Crewe's new strategy was to apply what he had developed through the electron-beam memory project to improve and diversify existing products. His experience with field emission technology brought Crewe back to his original work with electron microscopy. Crewe had successfully developed a miniaturized electron source that could be adapted to existing conventional microscopes, dramatically improving their performance. The ramifications of his invention were clear. Crewe's method could transform existing conventional microscopes to high-performance models at a fraction of the cost. Moreover, the sales volume of these modified microscopes was expected to grow with the target market's demand for more precise quality-control measures. By using Crewe's discovery, silicon-chip and other high-tech manufacturers—who have had to compromise precision production and quality control by using a mix of high- and low-performance microscopes—could have the capability to increase their production quality.

Crewe recognized the fact that he could achieve a sustained competitive advantage against the main producer of high-performance microscopes by offering cost-efficient innovative models to a broad market segment. This commercial potential and Crewe's desire to establish his own company persuaded the Boston-based Amray Corporation (which produces conventional microscopes) and Crosspoint Ventures to jointly fund Orchid One. While stewarding the product-oriented research and development efforts of the new company, Crewe has been permitted to cut back to a half-time term of appointment at the university. With assistance from the TCC, the new company was based at an incubator site near Crewe's residence and staffed by personnel selected by Crewe for their overlapping skills and nationwide contacts in the highly specialized areas of microscopy and field emission technology. The core staff included a computer programmer, engineers, and a technical and marketing manager with past experience in another university's microscopy department.

In 1988 the company was entering the final stages of prototype testing and product design work. Future plans included further development of industrial applications for electron-beam memory and related field-emission technology innovations.

Orchid One's overall strategy is to create value for customers by providing unique low-cost, customer-responsive, high-performance instruments. Rather than compete directly with existing models, the company plans to maintain its competitive advantages by creating strong new market niches—by "designing machines the competition has never conceived of," as Crewe himself has remarked.

Case History #2: The Traditional Model

Midway Labs, Incorporated—founded in 1986—is working to develop highly efficient, low-maintenance, and cost-competitive solar power systems. The company's first product exploits recent advances in optical theory developed by Dr. Roland Winston and Dr. Joseph O'Gallagher of the Physics Department at the University of Chicago.

Presently, photovoltaic systems rely on either low-cost, minimum-flux solar cells joined in flat panels without concentration, or else costly but highly efficient high-flux cells used in tracking concentrators. Recent innovations have led to the development of "one sun cells," which are dramatically more efficient than available systems.

Midway Labs has developed variants of these highly efficient cells, as well as a system to maximize their efficiency via nontracking optical concentrators. Through the use of TCC funds and their own assets, Midway Labs had developed a photovoltaic power source that offers a significant cost advantage and excellent reliability, and demands minimum maintenance.

The company's current operations are located on the university campus in close proximity to Drs. Winston and O'Gallagher, who serve as technical consultants to the company. The initial designs for applying their power source range from grid-connected residential use to remote-site electrical loads. The anticipated market for Midway's system encompasses the rural and remote parts of this country as well as regions in underdeveloped nations—wherever the need for electricity has increased but cannot be met by conventional power systems.

In this case, an outside business entity had acted through an informal network of referrals and identified a university-owned technology that complemented the company portfolio and performance targets. This method of technology transfer may be characterized as a process of reactive muddling-through, in which the institution relies on industry's general knowledge of its research activities—or on faculty members' industrial contacts—to commercialize its technology.

The process of commercialization in the traditional model may at times be more systematic than is indicated by the Midway Labs case. Some institutions maintain communication channels to relevant industries, and staff to introduce and market the institution's technology. The effective performance of this model is largely contingent on the institution's ability to manage its own technology portfolio.

Case History #3: The Institutional Model

The health-care industry has been in a state of flux since "capitated care" became a federal policy in 1983. Essentially, capitated care places a limit on reimbursement based on the category of medical services provided. In 1987, an estimated 60 percent of all care provided in hospitals was capitated. In effect, the new policy forces health-care providers to curtail costs and maintain the same standard of care.

The pressure for health-care institutions to introduce cost containment measures has recently been focused on areas such as preoperative testing that traditionally had not been evaluated in terms of cost. In 1984, Blue Cross and Blue Shield estimated that $30 billion was spent on preoperative testing and follow-up evaluation in the United States. The same study estimated that $12–18 billion could be saved by developing a more effective method of selecting preoperative testing. In addition to cost reduction pressures, the growing body of research examining the medical and legal risks associated with unnecessary medical tests raised the level of concern among health-care providers.

These prevailing concerns have created a receptive environment for health-care products that contain cost, increase efficiency, and reduce risk. A recently formed University of Chicago spin-off addresses health-care management and cost-containment problems that are commonly experienced by hospitals across the nation.

HealthQual Systems Corporation (HealthQual) was formed by ARCH in October 1987 to design, manufacture, and market a unique product line developed by Dr. Michael Roizen, chairman of the University of Chicago Department of Anesthesia and Critical Care at the University of Chicago Hospitals. As one of the university's first institutional spin-offs, HealthQual provides a recent example of how the university technology-transfer program operates.

In the spring of 1987—after an initial analysis of the project and a determination that it would indeed result in a viable spin-off—an ARCH associate (who would later become a principal manager of the new venture) was assigned to the project as coordinator of product develop-ment and marketing research. Simultaneously, ARCH initiated patent proceedings on both products and formed the new corporation, to which it granted exclusive license of the intellectual property relating to both products.

Under the direction of ARCH management, the associate worked closely with Dr. Roizen and technical consultants to develop working prototypes of the company's first two products. The TCC provided initial funding for the prototype development and assisted in preparation of a BIF application to the Illinois Department of Commerce and Community Affairs. These funds provided the necessary capital for first-stage proto-type development and initial concept-testing of the two products.

HealthQuiz, and the Chicago Druggist (CD), are the first two in a line of HealthQual products designed for the critical-care environment. Mis-placed drugs add to hospital costs and pose a health hazard in critical-care environments. Current systems cause a significant proportion of drugs to be misplaced in storage areas. The Chicago Druggist (CD) is an ampule storage system designed to be used where drugs must be segre-gated by drug type, readily located, and held securely.

HealthQuiz is a user friendly four-button computer with a removable memory cartridge that recommends appropriate tests for patients prior to surgery, based on the patients' responses to medical history questions. The software is contained on a single PROM (programmable read-only memory) chip, which will be replaced every six months to match the pace of medical advances. Plans for the development of HealthQuiz also include applications for emergency room testing, women's health, and corporate health programs. The system offers the following competitive advantages over the current medical practice of ordering complete batter-ies of preoperative tests: a reduction of preoperative test costs, medical risk to the patient, and the physician's and institution's medicolegal risk; improved maintenance of the patient's detailed medical history and profile for records; and an improved capability for tracking physician performance.

By mid-1988, third-stage prototypes of the two HealthQual products had been built and tested. A commitment of initial seed financing for HealthQual's expansion had been secured through the mutual efforts of Dr. Roizen and ARCH. Dr. Roizen would retain his position at the university and yet continue to be actively involved in development of the HealthQual product line. Dr. Roizen had also applied for a large grant to further his scientific validation of the HealthQual concepts.

Each of the spin-off case histories presented above has demonstrated the importance of an institution's having a clear concept of its commercialization policies, as well as the flexibility to apply or reinterpret the policies as institutional goals change. In the final section, a brief overview of common issues confronted by institutions in the process of establishing technology-transfer programs will be discussed. Additionally, the University of Chicago's method of managing internal resources through a centralized computer network is presented.

Policies to Support Technology Transfer

The preceding case histories serve to identify the rudimentary issues that institutions involved in technology commercialization must address. Foremost among these are the following:

1. Will faculty be diverted from their academic responsibilities through involvement in commercially oriented research?

2. Will corporate funding or research have a chilling effect on the tradition of knowledge transfer from the university to the public?

3. Should faculty have an entrepreneurial interest in a business that is supporting their research?

4. What investment risks and tax consequences can result from the university's potential equity position in spin-off ventures?

5. Should the university adopt a formal set of rules governing the interaction and acceptance of research support from nonconventional sources such as private investors or venture capitalists?

The question of how commercially oriented activity has changed the academic environment is now part of the recurrent background debate among and within institutions involved in technology transfer. The dwindling pool of federal funding and increased cost-of-product development in areas such as biotechnology (where federal regulatory requirements dictate high development costs) has spurred the growth of industry-sponsored research. Recent estimates of the amount of industry-sponsored academic research indicate that, in contrast to the relatively low levels of support for all other fields (3–4 percent), biotechnology research represents between 16 and 24 percent of industry funding (Blumenthal 1986).

Joint research agreements in this area are complicated by the fact that patentability is generally more complex for biotechnology products. Arguments for the reform of existing intellectual property law say the current law reinforces strategies that could increase the tension between academia's concern for uninhibited knowledge transfer and industry's concern for confidentiality.

> The more science advances, the more difficult it will be to meet the non-obviousness criterion if patent protection of research results is the norm. The more research is done secretly, however, the less information can be considered constructive knowledge and the easier it will be to meet the non-obviousness requirement. As a result, the requirement may serve as an incentive for industrial sponsors of joint research to push for more trade secret protection in certain basic areas of scientific inquiry. (Korn 1987)

Despite these complexities, tentative resolutions have been achieved by institutions through a process of case-by-case evaluation. As professionals engaged in technology transfer discuss both the real and perceived obstacles to developing programs at their institutions, a consensus may emerge that addresses the most prevalent concerns. Although each institution has a unique set of resources and culture, the most effective technology-transfer programs (in terms of active involvement with industry and profit) have successfully reconciled the academic concern for freedom, indemnification, and intellectual property ownership with the investment or industrial partner's requirement for confidentiality and financial return.

A policy environment that supports university–industry participation and involvement in state economic-development efforts such as the Illinois TCC program necessarily begins with the institution's intellectual property policy. Despite institutional differences, most universities with established technology-transfer programs have revised their intellectual property policy into a clear and formal statement that supports technology transfer and reflects the most recent federal patent law amendments. Such policies are also designed to protect and allow the institution to fulfill its obligation to the public sector by making research advances known and available to the public. Most universities in this category now offer incentives and rewards to faculty through the distribution of royalties. In conjunction with the revision of patent policy, institutions like the University of Chicago, Washington University, and the University of Utah have either established a patent management office or in some cases nonprofit or for-profit entities to manage the process of technology transfer.

The issue of how much the inventor's share should be in order to stimulate faculty involvement in a technology-transfer program varies among institutions, and to a large extent depends on the sources of support that fund the program in the first place. A recent poll of 12 universities indicated that most institutions use a royalty distribution formula allowing faculty to receive directly between 25 and 50 percent of the derived income, and eventually a portion is redistributed from the institutional share to the inventor's lab. The question of how much a faculty member should receive either directly as personal income or indirectly as laboratory income is resolved differently from institution to institution.

The majority of institutions deduct patent costs directly from royalty income received, before distributing the income internally. The more established technology-transfer programs—such as those at Stanford, MIT, and WARF—(the Wisconsin Alumni Research Foundation) usually require a licensee to pay a portion of the patent filing fee in addition to any up-front payments or royalties generated; and they manage their patent expenses through a self-sustaining budget. The director of the Stanford program has noted that during 1987 Stanford received approximately $6,100,000 from 116 cases. Only six of these required patent expenses exceeding $100,000, and 5 percent of the cases typically produce 75 percent of the income. In fact, most of the universities polled indicated that the majority of their cases bring in under $100,000. To compensate for that fact, institutions use a sliding scale of return, which provides the inventor and division with a larger share of the first $100,000 in income and a smaller share of all additional income generated by the patent.

The time required to build a program with operating budgets as large as those at MIT, WARF, or Stanford should not be underestimated. As

Derek Bok, president of Harvard University, has commented, "It takes 1,000 disclosures to yield 100 patents. It takes 100 patents to yield ten licenses. It takes ten licenses to yield one that earns over, say $50,000 a year" (Korn 1987, 193). As illustrated in Figure 11.3—the compared royalty distributions of three institutions with different patent management systems—the point at which patent costs are deducted (or not deducted) from the income generated by royalty clearly affects the amount remaining to be distributed to the faculty and institution.

In concert with the effort to clarify and amend their intellectual property policies, universities with established technology-transfer programs have developed methods to monitor the research resources of their institution. The promotion of a university's intellectual property portfolio begins with an internal compilation of faculty research interests, and results in the creation of a method for technology identification and assessment. At the University of Chicago, a centralized information network (UCAT) has been constructed to systematize the institution's internal inventory of research activities and allow screened access to industries interested in licensing a technology or entering into a collaborative research agreement.

A New Method for Managing Institutional Resources: UCAT

The University of Chicago Available Technologies Network (UCAT) catalogs and makes accessible by computer all the research interests of the faculty at the University of Chicago. The creation of a central data base is a vital step in any effort to encourage the exchange of information and technology between private industry and the university. Until now, information about faculty research, patents, copyrights, licensing agreements, and other significant information has been managed in a decentralized fashion throughout the university, rendering it nearly impossible—certainly prohibitively difficult—for an outside agency to locate research areas of common interest. With a large, highly active, and mobile faculty, it is easy for any given office or department within the university to overlook important projects and discoveries in progress on campus.

In combination with other methods of technology identification (e.g., through funding requests for prototype development funds, and word of mouth), the UCAT network will now allow the TCC and other university offices to synthesize effectively and share knowledge resulting from research at the University of Chicago. UCAT will function as an ongoing

Figure 11.3   Royalty Distribution at Stanford University, the University of Chicago, and Virginia Polytechnic Institute

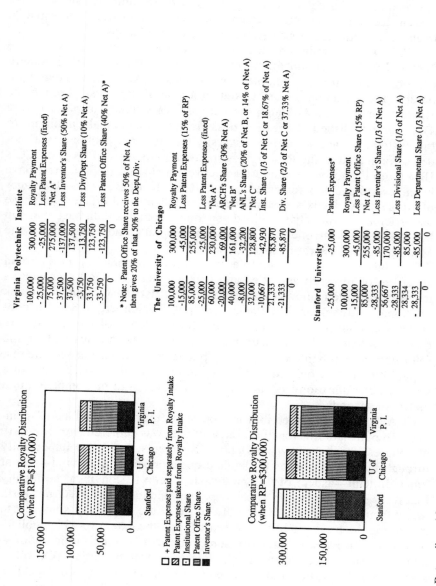

Virginia Polytechnic Institute

100,000	300,000	Royalty Payment
-25,000	-25,000	Less Patent Expenses (fixed)
75,000	275,000	"Net A"
-37,500	-137,000	Less Inventor's Share (50% Net A)
37,500	137,500	
-3,750	-13,750	Less Div/Dept Share (10% Net A)
33,750	123,750	
-33,750	-123,750	Less Patent Office Share (40% Net A)*
0	0	

* Note: Patent Office Share receives 50% of Net A,
then gives 20% of that 50% to the Dept./Div.

The University of Chicago

100,000	300,000	Royalty Payment
-15,000	-45,000	Less Patent Expenses (15% of RP)
85,000	255,000	
-25,000	-25,000	Less Patent Expenses (fixed)
60,000	230,000	"Net A"
-20,000	-69,000	ARCH's Share (30% Net A)
40,000	161,000	"Net B"
-8,000	-32,200	ANL's Share (20% of Net B, or 14% of Net A)
32,000	128,800	"Net C"
-10,667	-42,930	Inst. Share (1/3 of Net C or 18.67% of Net A)
21,333	85,870	
-21,333	-85,870	Div. Share (2/3 of Net C or 37.33% Net A)
0	0	

Stanford University

-25,000	-25,000	Patent Expenses*
100,000	300,000	Royalty Payment
-15,000	-45,000	Less Patent Office Share (15% RP)
85,000	255,000	"Net A"
-28,333	-85,000	Less Inventor's Share (1/3 of Net A)
56,667	170,000	
-28,333	-85,000	Less Divisional Share (1/3 of Net A)
28,334	85,000	
-28,333	-85,000	Less Departmental Share (1/3 Net A)
0	0	

Source: Data collected from University of Chicago Technology Commercialization Center survey of university royalty income distribution policies, 1988.

information resource mechanism to organize the internal stock of research activities and offer screened access to industries interested in licensing a technology or pursuing research partnerships.

UCAT was created using the Odesta Software Double Helix II. The Technology Commercialization Center and the Office of Research Administration share an office complex that utilizes 14 Apple/MacIntosh computers. These computers are linked to a central hard disk through the Appleshare system. Although the network was already in place before TCC was established, it has been possible to adapt and make additions to the system. By accessing the university's mainframe computer via modems connected to the Research Administration's Apple PCs and TCC's IBM (with the software to interface between IBM and Apple), there have been several avenues for creating the UCAT data base. The advantages of Apple's and Helix's system include simplicity and efficiency. These user-friendly features have helped generate a complex data base that can be accessed by anyone with even a rudimentary understanding of the MacIntosh PC's operating logic.

The university spent several years collecting information on its broad spectrum of research activities and technologies. The Helix program's icon-driven system allows the user to design a data base capable of quick relational searches and customized output (Hirschberg 1988). To construct a customized search and output form, chunks of information are assigned a field within the data base. In UCAT, these fields include the following: research interests, patents, publications, licensing agreements, copyrights, and associated faculty. Each entry field is coded through a keyword classification system that allows searches to be completed in a matter of seconds.

UCAT contains detailed information on more than 1,300 faculty members. To pull all of this together, relevant data was gathered from a variety of sources: from the divisions and departments, and most importantly through faculty responses to detailed questionnaires. One vital—and at times difficult—aspect of this process has been to convince the faculty that it is worth their while to participate in updating the UCAT network. Many faculty members are not fully aware of the institution's interest in marketing the products of their research, and the extent to which collaboration with private industry—or some other commercialization option—may be advantageous to them. The Research Administration/TCC office strives to make participation simple and to convey the value of UCAT to individual investigators by telling success stories as well as describing the possible intellectual and financial rewards of facilitated technology transfer. The TCC not only hopes effectively to target researchers who have knowledge and experience useful to industry, but also to help the faculty locate outside support for promoting their inventions. UCAT is a significant part of the TCC and ARCH's two-way tracking system to coordinate communication between the university and private enterprise.

UCAT is designed to contain large amounts of textual information that can be located by a range of different criteria. Each faculty member in the physical and biological sciences has an information sheet entered into the data bank. Queries can be designed to meet quickly the needs of industries and the university alike. The desired end product may vary from a simple list of faculty names, to detailed descriptions of current research activities of those faculty sharing some common interest, experimental methodology, or experience with certain instrumentation. Anything deemed to be a distinguishing indicator of these areas has been listed in the keyword field. Searches are executed by typing in a keyword and selecting the desired arrangement of faculty information assigned that keyword. If, for example, one wishes to locate all the faculty members who are studying solar energy, the key "solar" would be entered and "location" would be selected. In a few seconds, a list of names and addresses would be generated. Then one might narrow the list down—with a quick touch, or a wave of the wand—to the subset of, say, Fermi Lab physicists who answer to the keyword "solar." Crossing over then into the field of faculty names, one could call up descriptions of the individual researchers (Figure 11.4).

Since the TCC works closely with the Office of Research Administration—which is responsible for providing grant and funding services to the faculty—UCAT may very well be used to locate faculty members who have interests that correspond with the solicitation announcements that come in from both private and federal agencies. Just as important to the TCC as its role in commercializing existing research products and processes is its effort to work with the university's Office of Research Administration to identify ongoing faculty research that might require additional support to nurture its development and that may perhaps realize commercial value at a later stage. Promoting a relationship between lesser known funding programs and the faculty may help to diversify the research areas in which investigators enjoy support. For reference purposes, the funding/faculty data can easily be categorized and cross-listed by a number of relevant criteria, printed out by any type of office printer and strewn selectively about to generate ideas and spark contacts.

The Helix software allows for graphic, as well as textual and numerical, entries. With the new Mac II computer, it will be a simple matter to scan

Figure 11.4   A Typical Faculty Information Sheet in the UCAT Data Bank

Name:	XXXXX		
Title:	Professor	Title Code:	3.0
Office	Dept. of Physics	Mail Code:	03
Phone:	XXXXX	Name:	xxx-xxxx
Appointments:	Physics, Fermi Lab	Dept. Code:	8.0

Keywords: Solar; Particle; Energy Conversion;

Research:

I. Solar Energy. The new discipline of nonimaging optics makes possible solar energy concentrators that do not track the sun. Our group is developing collectors following these principles for the generation of high temperatures (up to 300 degrees Centigrade).
II. Experimental particle physics. The weak interactions of hyperons provide an important testing ground for theories of the symmetry and structure of elementary particles. Our group is analyzing a Fermilab experiment which made precision measurements of polarized sigma minus beta decay.

Patents:

Controlled Directional Scattering Cavity for Tubular Absorber; Compound Parabolic Concentrators with Cavity for Absorber (1/83); Energy Transmission with Respect at Convex Sources and Receivers (pending); Energy Transmission (12/80); Nonimaging Radiant Energy Directional Device (12/80)

Publications:

"Axially Symmetric Nonimaging Flux Concentrators with Maximum Theoretical Concentration Ration," Journal of the Optical Society of America (1987).

Source: The authors.

images into the data base. These graphics may at times communicate research pursuits and discoveries more effectively and meaningfully than short prose descriptions. The system will also be used by TCC at national and international technology-transfer fairs to showcase the university's available inventions both descriptively and graphically. By exploiting the benefits of this and other rapidly improving computer technologies, the TCC will be more efficient and dynamic in sponsoring communication and collaboration between university researchers and outside interests.

Conclusion

The events surrounding the creation of the University of Chicago's spin-off ventures highlight certain fundamental resources necessary to the successful establishment of a start-up. Any conclusions drawn from these histories must be somewhat limited due to their early stage of development. One thing is clear, however: Spin-offs require managers who can act in an ambiguous environment, make decisions based on technical knowledge of the product to be commercialized, and exhibit a keen market awareness. This is underscored by a recent survey in which ten high-tech CEOs identified the components of an effective management strategy for small-niche, high-tech companies. According to the interviewed CEOs, the two main determinants of success are:

1. Market orientation. Although many engineers and scientists are familiar enough with emerging technologies, the ability relevant to commercialization is to see the niches in the marketplace opened by these new technologies and come up with the needed products or services that people will buy in quantity. Selecting the right market niche for the technology is the key skill—and an essential ingredient for success in the high-tech world.

2. People. The dynamic environment of a high-technology firm requires people who understand the company's situation, objectives, and approach and will act promptly and decisively in situations of substantial uncertainty and risk (Lauenstein 1987).

Perhaps the most relevant contribution of the university's technology-transfer program has been its ability to complement the technical knowledge of faculty involved in the formation of spin-offs with its marketing and management expertise during the preliminary start-up phase of the companies. The university can act through its commercialization entities to reduce the spin-off mortality rate by providing competitive resources that are usually built up over time by a new venture—such as financing sources; reliable supplier/distributor contacts; and legal, accounting, and management consulting.

Given the embryonic state of the technology-transfer program at the University of Chicago, the empirical implications that may be derived from the development paths of its spin-offs are only now emerging. A positive risk–return relationship does seem to exist for institutions that are committed to long-term investment of resources. Over time, the university can be expected to develop strategies for coordinating the network of resources (financial and human capital) required to commercialize research results, reduce the start-up time of spin-offs, and spur the level of innovation in the institution. In this scenario,

> the process of innovation can be seen as circular, rather than sequential. Innovative ideas can spring up anywhere, and the multiplicity of sources increases the probability of positive outcome. A result of the process are the following two important characteristics: A high level of interpersonal contacts among players stimulates a higher joint perception of the complete business vision; Every new innovation adds up to an enlarged inventory of skills and knowledge which in turn increases the speed and improves the quality of future knowledge which in turn increases the speed and improves the quality of future innovations. Coordination mechanisms cannot be conventional but must be specific to the characteristics of the firms involved. (Lorenzoni and Ornati 1988)

This observation regarding the interaction between established firms and new ventures may also be applied to the exchanges between the established resources of a university and its spin-offs. The relationship between the faculty and the university's commercialization agencies is mutually beneficial, for it reinforces the institution's own innovative base by nurturing research in new technologies and thereby strengthens both its internal resources and external networks.

Notes

1. Estimated at about $2.5 billion in 1986.
2. The University of Chicago is the prime contractor for this DOE-supported laboratory.

References

AmeriTrust Corporation, *Choosing a Future—Steps to Revitalize the Mid-American Economy over the Next Decade,* 1986, p. 14–18.

David Blumenthal, Michael Gluck, Karen Seashore Louis, and David Wise, "Industrial Support of University Research in Biotechnology," *Science* 231 (1986):242.

Cullom Davis, "Illinois the Composite Economy," *Illinois Issues* (March 1988):20.

Irwin Feller, "Evaluating State Advanced Technology Programs," paper prepared for the U.S./European Conference on Regional Strategies for Innovation and Enterprise Competitiveness, Neuchâtel, Switzerland, September 24, 1987, p. 28.

Nancy E. Haar, Jenniffer Starr, and Ian C. Macmillan, "Informal Risk Capital Investors: Investment Patterns on the East Coast of the U.S.A.," *Journal of Business Venturing* 3(1988):11–29.

Gary Hirschberg, "Double Helix Takes Form," *MacUser* (May 1988):90.

David Korn, "Patent and Trade Secret Protection in University–Industry Research Relationships in Biotechnology," *Harvard Journal on Legislation* 24, 117(1987):217.

Milton C. Lauenstein, Daniel J. McCarthy, and Francis C. Spital, "Managing Growth at High Technology Companies: A View from the Top," *Academy of Management Executive* 1, 3(1987):320.

Hianni Lorenzoni and Oscar A. Ornati, "Constellations of Firms and New Ventures," *Journal of Business Venturing* 3(1988):52.

Pamela Sherrod, "Academics, High-Tech Small Firms Link Up," *Chicago Tribune*, Business Section, October 12, 1987.

12

How University Research Results Become a Business: The Case of the University of Connecticut

Ilze Krisst

The University of Connecticut—like many other research universities—is being challenged by a confluence of events to decide not *whether* it will nurture the transfer of research results into the economy, but rather *how well* it will do it. Isaac Bashevis Singer has a story in which he describes a moment of intimacy between a villager and his wife: "She pretended she was asleep, and he pretended that he did not know what he was doing." We have concluded that the only prudent course is to recognize and indeed to facilitate the relationships between the academic and the business world and to legitimize the spin-off results from this union.

This chapter considers the context in which this process evolved, as well as certain specific issues that the University of Connecticut has identified from the experience of the past four years.

Background

The modern, American university—born in 1636—has been incredibly successful and creative and is considered to be one of the most vital strategic resources in the United States. The country has long been a leader in producing Nobel prizes, advanced science and technology, and innovative start-up companies.

Many of the spectacular technological advances that occurred during World War II originated in university laboratories. Subsequently, federal funding for university research increased significantly. This support signaled the beginning of an era of unsurpassed productivity for the U.S. research university. Today, the federal government continues to have a significant impact via the approximately $55 billion that it spends on research and development, as well as through the $10 billion spent directly on campus. Further, although the proportion of industrial research and development support in the leading research universities rose in the mid-1980s by 38 percent, the federal government remains the dominant patron, providing 64.66 percent of the funds. In addition, recent legislation has encouraged the disclosure and utilization of university-owned research results, based on the premise that the public has a legitimate right to interim benefits from the results of government-funded research and also in order to strengthen the competitive position of the United States in the international market.

Finally, the federal government can provide one of the most critical ingredients to the university research process—namely, stability. It does this via long-term funding, and by certain economic policies designed to encourage the private sector to invest in academic research and development. Any significant hiatus in funding for research always has a seriously deleterious effect on the ensuing product. Equipment becomes obsolete; ideas wither or find other channels of productivity. As the locus of public policy formulation, the federal government must ever make it clear that the country's solid research base does not consist of a series of intermittent projects, but rather is a process that occurs over a continuum of time.

Clearly, universities in the United States have also enjoyed generous support from private sources. Many major research universities were founded on private industrial fortunes, and it is not uncommon for industrialists to sit on the governing boards of both private and public institutions. Today, corporate support for academic research has shifted from passive donations, fellowships, and grants to a relationship that is often more like a partnership. These partnerships include traditional modes such as consulting and seminars as well as new ones such as cooperative research projects, a variety of consortia, and more recently a focus on the creation of new businesses. Universities are actively seeking additional funding for research from industry. Studies indicate that in the period between 1980 and 1985 industrial research support rose by 30 percent, from 3.59 percent to 4.95 percent.

Moreover, a July 1987 report of the National Governors Association urges these leaders to expand greatly the use of their public university system in encouraging economic growth, and business is actively promoting the synergy created by interaction with universities. Says John Rydz of Emhart Corporation, "We're becoming a little looser in our willingness to open up. We're going to have to . . . and we're becoming more willing to pay royalties on patents, to pay royalties to universities" (private communication). A variety of factors have contributed to this trend: changes in federal laws relating to research done at universities; the challenge to restore the competitive position of the United States in the global marketplace; the need to expand the funding base for research projects; and the growing entrepreneurial interest of faculty members.

Before turning to the particular process and mechanisms employed by the University of Connecticut in its plunge into the technology-transfer arena, therefore, one must recognize that its efforts in this regard are part of a trend that has involved considerable local and national introspection and speculation about the role and mission of academic institutions. The writer of a provocative letter published in the January 29, 1988, issue of *Science* argues that the nation's changing values and the associated tyranny of the bottom line have reduced industry's incentives to take risks and have, in fact, weakened the ability of the United States to compete in the world marketplace. The letter writer is unequivocal about the absolute need to transfer technology to the industrial system where it will benefit the nation. However, he also contends that research universities are engaged in science and engineering; they are not engaged in technology. While universities do have a great wealth of knowledge and cutting-edge research results to transfer, technology is the application of this knowledge. Though universities may strive for better methods to transfer it, industry will always have the primary responsibility to develop and perfect technology. University involvement in the development of spin-off companies and similar endeavors is probably most appropriately viewed as an important ancillary, rather than primary, activity.

Furthermore, even among those separatists who take the position that "the business of America is business" and the business of universities is to educate, there appears to be a consensus these days that what universities feel they do best—basic research—does relate in some admittedly ill-defined way to the innovation that is vital to the future of U.S. industry. University laboratories have historically been the source of people on whose ideas such innovation is based. Whatever the form of interaction between our campuses and the "real" world—be it consultantship, sponsored research program, or spin off—the relationship enhances investigator initiative, stimulates an exchange of ideas, and expands the opportunity for collaborative arrangements in which both sides contribute their strengths and also have their needs met. However, it cannot be emphasized too much that the catalyst in these alliances will probably always be the one-to-one relationships established between creative scientists and engineers on opposite sides of the fence.

Policy Issues

Linkages with business and industry are expected to grow at the University of Connecticut not only in response to national trends, but also as a result of the development of the Connecticut Technology Park adjacent to the campus, as well as certain new incentives from state government.

In the summer of 1988, the university's Policy Advisory Committee on University–Industry Relations completed a year-long study of issues relating to faculty and staff members' relationships with business and industry. The process culminated in a statement of policy. The committee considered various topics including the appropriateness of such working relationships for faculty members, and the extent of them; the appropriateness of certain outside business ventures, and the time that can be spent on them; and faculty rights to royalty income from the commercialization of their inventions.

The guiding principle on these matters at the University of Connecticut is a board of trustees policy explicitly encouraging university–industry relationships. Controversy may still arise, however, when such activity conflicts with certain tenants of the university by-laws—namely, the principles of "carrying out scholarly work in a free and open atmosphere and to publish freely" and a faculty member's responsibility to "choose and pursue professional activities in a way that advances both institutional goals and professional development."

While the university is cognizant of the fact that patent rights and confidentiality are thorny issues for industry, it nonetheless insists that no research project be undertaken if it censors the final results, and that special care be taken with respect to work done by students. A great deal of flexibility and understanding by both sides is required to reconcile the legitimate need to protect industry's market position and the obligation of the university to disseminate information freely. The relationship may become intrinsically more complex as faculty members become consultants to companies spun off from research that they are still pursuing as part of an academic program.

In order to understand the question of how much energy and time a faculty member may devote to external activities, one must remember that his or her obligation to the university includes time spent in self-directed activity. This is not to be confused with private time—the time clearly not due the university. Consequently, all self-directed activity must be appropriate to professional standards and advance institutional goals. In addition, faculty members—as "officers of the university"—must be aware that they function as representatives of the institution and members of the academic profession. They are expected to meet a certain standard of impartiality and credibility. Thus it is incumbent on the faculty member to eschew even the appearance of conflict of interest, either in terms of commitment or financial gain. If faculty members' commitment to an outside activity is no longer appropriate to and compatible with their obligation to the university, a partial or full leave of absence may be requested. As increasing numbers of faculty members contemplate more extensive involvement in assorted business ventures, it will become necessary to devise ways to make such arrangements attractive by protecting the person's position and tenure status during the period of absence.

With respect to faculty members' right to royalty income, in the words of Anthony DiBenedetto—the former vice-president of academic affairs at the University of Connecticut—"One has the right to whatever the employer allows and the law does not limit" (private communication). According to Connecticut law, the university owns all inventions conceived at the university, and the inventor is entitled to a minimum of 20 percent of the royalty income received by the university. However, the law also gives to the board of trustees the discretionary power to increase this amount to 50 percent. Currently, proposals for a fixed revenue sharing scheme within this range are being examined by a subcommittee of the university's Policy Advisory Committee in order to make the process less cumbersome to administer and more objective than individually negotiated rates.

Finally, as the university begins to deal with an ever-increasing number of spin-off companies, it will also need to formulate and articulate a consistent position on faculty equity shares in start-up companies. Currently, neither state law, the university by-laws, nor university policy speak to the issue.

One of the most serious issues that will have to be reviewed is rooted in product liability law. The university is aware that it now risks being considered a participant in the "stream of commerce" as a result of its more aggressive position in the technology-transfer process. As court decisions establish new precedents, the universities are becoming increasingly more liable and more financially exposed, as they do not have the fund streams for major litigation and damage awards. Should the risk continue to increase, this may be one issue that will force universities seriously to reconsider their involvement in technology transfer.

Technology-Transfer Mechanisms

It should be fairly evident that U.S. universities are no newcomers to the transferring of research results into the economy. What is new—and this is certainly the case at the University of Connecticut—is that the transfer occurs no longer as a random, isolated event, but rather as an institutional phenomenon. Also, the university has now accepted the challenge of finding ways to assist gifted scientists with commercial ideas. Lacking an outlet for their entrepreneurial ideas, such individuals will remain frustrated, go to work in industry, or set up moonlighting operations with or without the knowledge of the institution. None of these alternatives do the parent institution much good.

According to state law, inventions developed by faculty and staff researchers are owned by the University of Connecticut and assigned to the University of Connecticut Research Foundation. The Research Foundation, by law, is the custodian of university research funds and intellectual property and has historically carried out the administrative function related to invention disclosures and patent applications. It can exercise discretionary power to assign intellectual property rights to other entities, which in turn may develop the commercial potential of such property through licensing arrangements, ventures, spin-offs, and other appropriate vehicles. By tradition, the Research Foundation itself has not engaged in overt and active marketing of intellectual properties nor in efforts to incubate new businesses. The companies previously engaged by the university for marketing purposes had been passive in this role.

Acting on recommendations formulated by the Policy Advisory Committee, the university established an Office of Cooperative Research and an Office of Technology Transfer. The Office of Cooperative Research promotes interaction between the university and industry through seminars, conferences, and other similar activities; formulates policy; and disseminates appropriate information to faculty and industry on the premise that positive attitudes will bring down the barriers to collaboration. The Office of Technology Transfer deals primarily with reviewing, processing, and determining the commercial potential of university invention disclosures and generally advises on all functions related to patents and licensing. It is funded through a recently consummated arrangement with University Technology Corporation (UTC), a national company created specifically to promote the transfer of intellectual innovations of universities to the appropriate companies. In 1985 the university entered into a contract with UTC, designating the latter to be its exclusive licensing agent. The contract specifies that income distribution between the university and UTC will be fifty-fifty. The university then shares its half with the inventor.

At about the same time as the university was entering into its association with UTC, the entrepreneurial then–assistant vice-president for research at the University Health Center spotted an opportunity created by the Research Foundation's reluctant posture toward commercialization of novel research results. In 1984 the vice-president established the University of Connecticut Research and Development Corporation (R&D Corp.), which operates at arm's length from the university and differs from UTC in that it concentrates on securing venture capital for start-up companies. In addition, it offers opportunities for the faculty inventor to acquire minority ownership positions in newly formed companies. The founder of R&D Corp.—who holds a Ph.D. in physiology—had excellent entrée to the university faculty; he argued that discoveries spring from laboratory research in the most unlikely situations and that the inventor frequently has no real taste for the business aspects of commercialization.

Approved by the university's board of trustees as a universitywide entity, the corporation was established to promote efficient transfer of technology and to support and encourage scientific research at the university and all its campuses including the Health Center. It is a wholly owned subsidiary of the University of Connecticut Foundation, which is a nonprofit private corporation—separate from the University of Connecticut Research Foundation—and the sole owner of R&D Corp. stock. The R&D Corp was modeled on similar entities that have been springing up throughout the country.

Specifically, it evaluates technologies, selects those with significant economic potential, assists in completing research and development for commercial viability, and arranges appropriate commercialization. R&D Corp. has the right of first refusal to develop disclosed techniques or products. If this right is waived, such development will be considered— per contract—by UTC. R&D Corp. is also empowered to negotiate the transfer—subsequent to assignment by the Research Foundation—of university research results to commercial partners and to return any income beyond the operating needs of the corporation to the Research Foundation in order to fund and support additional research.

Actually, the original servicing agreement between the R&D Corp. and the university specified that 75 percent of the net cash income (income less expenses) would go to the University Research Foundation, and 25 percent to the R&D Corp. Because the original servicing agreement failed to provide adequate operating revenue to the R&D Corp., another income distribution agreement was structured that increased the university's share to 85 percent, but also permitted the R&D Corp. to retain the first $250,000 of net cash income each fiscal year. This agreement raised many difficult issues since it placed a disproportionate burden for funding of the corporation on the inventors of the first several properties promoted by the R&D Corp. By law—as previously noted—the university shares its royalty income with the inventor. Even though on its face (and assuming an ample royalty flow) the R&D Corp. agreement is favorable to the university and the inventor, no income has been realized to date because of the provision allowing the R&D Corp. to retain $250,000 annually. At this time it is clear that the different perspectives of the inventor, the university, and the newly formed R&D Corp. may not have been very carefully considered at the initial stages. Inevitable as this may be in breaking new ground, some very important and generalizable questions have been raised by the situation.

As for the perspective of the R&D Corp., its founder says, "We are in this for the long haul." The corporation's income depends on the success of its venture projects. Its only operating fund is money off deals—including cash flow generated by royalties, servicing agreements, and equity shares.

Risk Capital

Working capital for R&D Corp. projects to set up new companies is derived from strategic commercial participants, private investors, debt financing, and venture capital.

"Concomitant with the growing awareness of universities as a source of new technology has been recognition of the need for greater availability of risk capital for new companies," says the corporation's president. Many states, including Connecticut, have encouraged the development of early-stage venture funds in order to promote the growth of high-technol-

ogy companies. The R&D Corp. president also points out that understanding the relationship of risk capital to technological innovation requires an awareness of the innovation process itself. Frequently this can
be lengthy—from seven to ten years—and it consists of three essential
stages: (1) invention; (2) transformation into new product or process; and
(3) commercialization. The second stage represents the time of greatest
risk as well as cost.

To minimize the risk, states frequently encourage participation by the
private sector. Sometimes this participation amounts to managing the
investments made with state funds; in some cases, however, the firm is
both investor and manager, with involvement from the state. State programs may take many forms, generally including some of the following:
(1) direct equity ownership; (2) royalty sharing; (3) research-and-development grants; (4) unsecured long-term debt; and (5) equity guarantees.
In most state programs, the enabling legislation specifies the criteria.

In Connecticut, efforts to foster the development of university-gener-
ated technologies are assisted by various state programs—including a
new Connecticut Seed Venture Fund, the Connecticut Product Development Corporation, and the Small Business Research Grant Programs.

Spin-off Companies

In discussing the genesis of the university's more aggressive role in the
commercialization process, it is appropriate to consider specific examples
from between 1984 and 1988 among the first companies founded on the
research results of University of Connecticut scientists. The common
denominator in all these situations is that the inventors were primarily
interested in their research and wished to remain in the university
environment. The point of view of these scientists has ranged from simply
not wanting to be bothered with the business aspects of commercialization, to wishing to be aware and involved as consultants, to actually
wanting hands-on operating responsibility.

The early 1980s had been a time of heavy venture-capital flow into
development of biomedical technology. Many faculty members, however,
were generally experiencing frustration because of their inability to
access venture capital and other innovative ways of attracting research
funds. In addition, numerous academic scientists were concerned about
the lack of adequate laboratory space for their research and about issues
relating to the time they were devoting to and the income they had
generated by consulting as well as the possibility of holding equity shares

in companies spun off from their work. Few, however, had any compelling interest in commercialization or in leaving academia.

Example #1

A University of Connecticut Medical School marine biologist had—as part of his routine research projects—isolated the chemical compound of the adhesive used by mussels to adhere to any surface underwater. In 1984 a Connecticut company was born from this discovery, on handshakes and promises. It was R&D Corp.'s first hot project.

The evolution of this particular company was a series of coincidences and unique circumstances. The inventor had given a paper—that is, made a public disclosure—of his invention without being aware that this action also set the clock ticking on the time limit for patent application. After being alerted to this fact by a member of his graduate student's family, the faculty inventor hired his own lawyer to file a patent application—apparently having lost faith in the willingness or ability of the university to act in a timely and efficient manner. Then it was the same graduate student's husband—understanding enough about the research to expect "mussel glue" to be commercializable—who went to the appropriate university administrator—who, at the time, was the current R&D Corp. president—to discuss licensing the property. In 1985—after some contentions pertaining to exclusivity—R&D Corp. formally executed a license to the new company, giving it exclusive rights to the chemical compound for the adhesive—which today is sold to the biomedical research market. The inventor's patent application was filed the day before the last handshake.

All these actions had been preceded by an agreement through which the inventor assigned his invention to the university, which in turn ceded the rights to the R&D Corp. However, the R&D Corp. continued to struggle with capitalization, because its agreement with the new company—which assured the licensor a royalty income—was structured in such a way that the actual flow would not start until the end of the third year. In fact, payment started after 18 months.

The technology had been successfully transferred, so to speak. R&D Corp. had launched its first project. After three and a half years of operation, the new company moved to new quarters; and according to its president, it was "turning away investors."

The process may have taken a different turn entirely, had it not been for that graduate student. Since it was the faculty inventor's prime objective to continue his research in an academic setting, it was the graduate student who became the conduit by which the knowledge and expertise were transferred to the new company. The inventor, who holds an equity position per the agreement with R&D Corp., currently serves as the company's consultant. Clearly—in this example as in every case—if expectations and perspectives of all the relevant parties are addressed up front, serious disagreements are far less likely to arise.

Example #2

More recently, the research and consulting activity of a professor of civil engineering resulted in a method—known as induced soil venting—used to clean up petroleum spills caused by leaking underground gasoline tanks. When the technology was disclosed to the university, it was deemed unmarketable and immediately released to the inventor. This eliminated any contention over royalty splits with the university. No attempt was made to apply for a patent.

This particular inventor, however, had a keen interest in exploring the marketability of his invention and in exercising a latent desire to savor the world of business. Lacking the time and necessary savvy to establish a company, he concluded early on that he needed to hire an attorney to protect his interests, and subsequently entered into negotiations with the R&D Corp. During 1987–88, R&D Corp. contributed to development of the new company by generating a business plan and securing leads for venture capital. Its deal with the inventor guarantees him a generous equity position in the company.

The new company is constituted of a president and two M.B.A.-level unpaid consultants. In late 1988, it was experiencing a strong dose of reality: the search for risk capital; the resignation of its president; and the October 19, 1988 stock market crash. The inventor—whose optimism is incontestable—expects to devote approximately 30 days a year to the company as a senior technical consultant. He contends that transferring the results of his research to the application state is necessary and appropriate, and that his presence in the company as a consultant lends it credibility and enhances its chances of success.

Example #3

A professor of electrical engineering who embodies a unique combination of both impressive business and academic credentials recently identified a need for instrumentation that would pinpoint incipient deficiencies in light voltage cables while they are in service. With the help of colleagues, a method to accomplish this was conceived and tested successful in the lab. A literature search indicated that no one else had a similar device.

At this same time, the inventor happened to be serving on the university's Policy Advisory Committee, which—in his case—engendered a new interest in forming his own company. Various other vehicles (including the R&D Corp.) were considered, but—having taken matters into his own hands before—he decided to go it alone. As of this writing, the final go is contingent on the outcome of the patent application.

Since he wants to have actual hands-on responsibility for the operation of his company, our faculty inventor is aware that during the first year the business will require 100-percent commitment, making it impossible for him to discharge his obligations to the university. His request for a sabbatical has been turned down, however, because starting a business is not considered to be an appropriate sabbatical activity. The professor is quite puzzled. Why can you not make a profit and grow intellectually at the same time? he wonders. A leave of absence is probably his only viable solution. And when he is ready to return to academic duties, he expects to hold the position of chairman of the board of the new venture— a position that will not require his daily presence.

Further, the professor argues that this type of activity benefits the university, students, and society at large. That is, real-life problems generate research funding and also ideas for significant basic-research projects; they provide students with the opportunity to apply knowledge; and research commercialization benefits society by creating products that perform better and are more cost effective.

Conclusion

The process by which technology is transferred from the academic arena into the economy cannot be separated from the structure and mission of the university. The federal government will continue to be looked to for enlightened public-policy formulation; universities will continue to focus on research and its results as an intrinsic component of their mission to educate students, create new knowledge, and produce the next generation of scholars and scientists; and industry will keep on searching for emerging technologies.

Creative, innovative scientists have had successful relationships with business and industry for decades. Whatever their motivation in doing so—be it a desire to access additional research funds or personal monetary gain; to see the practical application of their work or to contribute to the operation of a business—as individuals they are nevertheless still very interested in staying with the academic setting. It is clearly in the university's best interest to create an environment that encourages these faculty members—who frequently are the most creative and productive ones—to remain at the institution. The integrity of the institution itself as well as the rights of its faculty members need to be simultaneously protected. Certain obstacles—such as struggles over royalty distribution, equity positions, appropriate opportunities for consulting, and leaves of absence—will have to be overcome. It is to the advantage of university administrations to make the technology-transfer process work smoothly for all. By-laws and guidelines should not become constraints, but rather provide a structure that will allow both sides of the transfer to flourish.

Readings

Ted Bartalotta, "Taking Science to Market," *University of Connecticut Health Center Quarterly* 3, (Winter 1987–88):7–11.

Edward E. David, Jr., "Looking Ahead at University–Industry Partnerships," *Engineering Center Network Annual Proceedings* (1986).

Robert S. Friedman and Renee C. Friedman, *Sponsorship, Organization and Program Change at 100 Universities,* Pennsylvania State University, June 1988.

Government–University–Industry Research Roundtable, *New Alliances and Partnerships in American Science and Engineering* (Washington, D.C.: National Academy Press, 1986).

Lyle Hohnke, "Marketing New Technology from University Laboratories, and the Financing of New Ventures in the Capital Markets," *State of Connecticut Securities and Investment Division Bulletin* 3, 1 (November 1987).

"Learning the Ropes about Cable Insulation Reliability," *Scope* 20, 1, (February 1988).

"Letters," *Science* 239 (January 29, 1988).

Edward L. MacCordy, "Introduction Issues in the University–Industry Partnership," *Research Management Review* 1, 2.

State of Connecticut, *The General Statutes of Connecticut,* vols. 1 and 2, rev. January 1985.

University of Connecticut, *The University of Connecticut Laws and By-Laws,* Twelfth ed., 1985.

Robert Varrin and Diane Kukich, "Guidelines for Industry-sponsored Research at Universities," *Science* 227 (January 25, 1985).

13

Entrepreneurship at Purdue University

Stanley T. Thompson

Over the past decade, the United States has changed its business style because of tougher competition from Europe and Asia. One competitive advantage America has traditionally held over the rest of the world is technology. To remain competitive, its technological advancements have had to increase substantially. Major breeding grounds for technology are the research activities occurring at universities. How competitive the United States remains in the world's economy will depend in part on how quickly new technology is developed and commercialized at its universities.

Purdue University has—through the years—received millions of dollars in research funding and has generated volumes of research information. The university has educated and trained thousands of engineers and scientists for careers in industry. These graduates have taken with them research ideas and implemented those ideas in the marketplace. In the past, faculty and staff were not encouraged to commercialize their research activities. A faculty member's responsibilities were to educate, research, and publish—not to become a businessperson. Outside consulting activities were tolerated, but being an entrepreneur was a negative detail when seeking tenure.

In the early 1980s the State of Indiana began an initiative to turn its aging and rusting economy into a competitive and technology-oriented economy. The state established three nonprofit corporations to implement its strategic plan. The Corporation for Science and Technology was formed to fund development efforts in targeted high-technology industries. The Corporation for Innovation and Development was set up to provide follow-on funds for marketing developed products. Finally, the Institute for New Business Ventures was organized to offer educational conferences and counseling for business people in the state.

Purdue University was looked upon by the state to provide assistance in the area of technology. The governor wanted Purdue to create an Indiana version of Silicon Valley. To this end, two groups were formed to assist business people in accessing the university's resources. The Business and Industrial Development Center matches a company's requirement with the university resource needed to assist the company. The Technical Assistance Program provides a faculty member and a graduate student to work on the particular problems facing a specific business, using existing technologies.

The third form of assistance provided by the university was in the form of a high-technology business incubator: INventure. The incubator assists Purdue faculty and staff in commercializing their technology ideas by starting a company. The technology is licensed from Purdue and developed into a product that is sold through the spin-off company. INventure supports the early development of the spin-off company and improves the probability of the business still being in existence after five years—by 80 percent!

INventure is a for-profit high technology business incubator supported by Purdue University. Its mission is to achieve the following four objectives:

1. To be the focal point for high-technology corporate start-ups in the State of Indiana;

2. To establish a reputation within Indiana that attracts the financial resources needed for new venture capitalization;

3. To be a major source for entrepreneurial education and business guidance; and

4. To provide conditions and support systems to help ensure the development of start-up businesses.

INventure provides facilities and services to start-up companies for an average of two years. Facilities include office space, furniture, telephones, conference rooms, audio-visual equipment, personal computers, printers, software, a copier, and a postage machine. Services include a receptionist, a secretary, word processing, accounting, business plan writing, research proposal preparation, and business counsel on technical, marketing, financial, and administrative issues. A small amount of seed capital is also provided. The value of INventure's facilities and services over the two-year period ranges from $100,000 to $150,000. The start-up companies pay no fees; rather, INventure receives equity in the company. Typically, INventure receives a minority position of 20–30 percent of the initially issued stock. The entrepreneur(s) own the remaining 70–80 percent. INventure takes equity instead of cash because start-up businesses are cash poor and equity rich.

INventure has three potential markets for entrepreneurs. The primary market is among the faculty and staff members at Purdue. A secondary market is among entrepreneurs who locate near Purdue to tap resources at the university. Finally, the third market is any individual within Indiana or any other state who wants to start a high-tech company.

In two years of operation, INventure has incubated five companies. Three of the companies are spin-offs from Purdue, and the fourth business located in West Lafayette to access Purdue resources. The fifth company moved from Illinois to the incubator because of the services and resources offered by INventure. More than 120 entrepreneurs had been screened to arrive at the five incubated companies. (For each particular company, the screening process lasted anywhere from three to six months.) The first start-up company moved out of the incubator after 17 months.

Four for-profit corporations are equal shareholders in INventure. They are André Financial Corporation; Bioanalytical Systems, Incorporated; McClure Park, Incorporated; and Pritsker and Associates, Incorporated. McClure Park is a wholly owned corporation of the Purdue Research Foundation and represents Purdue's involvement in the incubator. The other three were founded by local businessmen who were at one time associated with Purdue.

The founding corporations are represented by individuals who comprise the directors and officers of the INventure. Each contributes a minimum of two days per month to assisting the incubator companies. The business and academic backgrounds of these individuals provides an example of the multidisciplinary nature of INventure. The president of INventure is also chairman of Pritsker and Associates. He is an adjunct professor of industrial engineering at Purdue and is responsible for the incubator's daily operations. The vice-president of INventure is also the vice-president of research at Purdue. Besides being an administrator, he is a professor of chemical engineering. Another vice-president of INventure is president of Bioanalytical Systems, as well as a part-time professor of chemistry. He is responsible for providing marketing and manufacturing assistance to the incubator. The treasurer of INventure is president of André Financial Corporation. He is responsible for providing financial, investment, and legal guidance to the incubator companies. The treasurer is a former chemistry professor at Purdue and was chief financial officer of Great Lakes Chemical before starting his own business.

Each of the officers actively participate in the management of the incubator companies. One INventure officer sits as a director on each company's board. The remaining three officers assist the company when requested. Since these are corporate investments, the staffs and resources of the founding companies are available to the incubator companies. The staffs provide technical, marketing, manufacturing, and administrative assistance.

Purdue and INventure

In the past, Purdue promoted the development of new technologies at the university, but did not encourage the spin off of companies. However, when approached by the state to provide leadership in technology, the university reversed its position and began encouraging spin-off corporations. INventure was established to assist and increase the likelihood of spin-off successes. There are several other reasons why Purdue founded INventure:

1. To commercialize the technology and generate license revenues to the university;

2. To retain graduating engineers and scientists in Indiana by creating challenging job opportunities;

3. To offer a diversity of employment opportunities to faculty members' spouses;

4. To provide real-world experiences to graduate and undergraduate students;

5. To expand the Purdue Research Park with technology-oriented companies; and

6. To diversify the local economy—which reduces the local effects of a national recession.

Until a few years ago Purdue had never licensed any technology from the university for more than $100,000. Its strategy has changed dramatically and is evident by the packaging of a $6-million licensing arrangement last year.

When a Purdue faculty or staff member develops a new technology, the inventor discloses his or her activity to the Office of Patents and Copyrights, where a brief form describing the idea is filed. The disclosure form is reviewed for any outside activity (licensing, consulting, or starting a business). Any technology developed in a faculty member's area of expertise is owned by Purdue, even if developed outside the university.

The second step in the licensing process is a university review of the technology to decide whether or not to protect it. If the technology is promising, Purdue Research Foundation either patents or copyrights it. If it seems to have a commercial market, a licensing agreement is negotiated. After all the legal expenses have been reimbursed, any royalties from the license are split equally between Purdue Research Foundation, the faculty member's department, and the faculty member.

Purdue's policy is to allow its faculty one day per week for outside activities. The activity can be teaching, consulting, or managing a business. Conflicts of interest are not a problem so long as a disclosure form is filed with the University Office of Patents and Copyrights. Indeed, such situations of potential conflict are informally encouraged because they typically result in a win-win solution. Both the university and the spin-off company usually benefit.

Case Studies

Case Example #1

Dr. Peter Kissinger started Bioanalytical Systems, Incorporated (BAS) before 1985—when the disclosure and licensing process was more difficult and entrepreneurship was rather discouraged at Purdue. Professors were supposed to teach, research, publish, and possibly consult with private industry. They were not to become business people.

Prior to coming to Purdue, Dr. Kissinger had begun research on instrumentation for the determination of trace amounts of chemicals in biological, environmental and industrial materials. Then in 1975 Dr. Kissinger founded BAS with two other colleagues while sitting around a kitchen table. In the evenings and on weekends, the three developed academic concepts into what they believed could be commercially viable products. By 1977 BAS occupied a three-car garage and most of Kissinger's house. In 1978 the firm purchased a building in Purdue Research Park. In 1982 that building was doubled in size, and an additional 50,000-square-foot production facility was added in 1983.

At present, BAS markets complete analytical systems in the $5,000–55,000 range. BAS maintains expertise in chromatographic and electrochemical instrumentation for analytical purposes or for more fundamental research on the electrochemical behavior of organic, inorganic, and biochemical systems. Its objective is to maintain and increase its market share by providing quality products and support. Sales operations are located in Atlanta, San Francisco, Dallas, Chicago, Columbus, Central New Jersey, Boston, and Washington D.C. International sales account for 30–35 percent of total revenues. Current employment is 75 individuals worldwide.

Case Example #2

Dr. Pritsker was a pioneer in the field of simulation at Arizona State University as well as at Purdue. He developed the SLAM II simulation language, which is the leading software in the industry. In 1973 he formed Pritsker and Associates, Incorporated (P&A) with two of his industrial engineering graduate students. Dr. Pritsker remained a full-time professor for several years before joining the business full-time in 1980. During these years, his involvement with his own company was discouraged by Purdue University.

P&A began as a consulting firm that specialized in the simulation of problems in the automotive, aerospace, electronics, heavy industrial, and packaged goods industries. From a consulting foundation, P&A has grown into a company that provides software products, training, and support. Employment has increased modestly over the years to 80 people. P&A has more than 3,000 installations worldwide with offices in West Lafayette, Indianapolis, and California.

An often overlooked benefit of university spin-offs is its second generation of spin-off businesses. In 1985 P&A formed Factrol, a corporation consisting of P&A employees, to develop shop-floor scheduling software.

Factrol was nurtured with P&A's facilities for two years before moving into its own offices next door. After three years, Factrol employs 25 people, and its revenues are 33 percent of P&A's. Most of the employees are Purdue graduates.

The next two case examples were started after Purdue's change in philosophy around 1984. Both companies were blessed and encouraged by the university.

Case Example #3

Batch Process Technologies, Incorporated (BP Tech) was cofounded by Dr. Gintaras Reklaitis, Dr. Martin Okos, and Dr. Girish Joglekar. An NSF/IUC (Industry–University Cooperation) grant had funded their research work concerning the feasibility of simulating semicontinuous or batch processes. Drs. Reklaitis and Okos were the principal investigators. Dr. Joglekar was a postdoctorate student at the time, and he headed the research team. After the in-university project was completed, BP Tech was formed in 1984, and development funding sought.

In 1985 a consortium of ten companies funded a two-year BP Tech project that produced a software product called BATCHES, which simulates semicontinuous and batch processes. Dr. Joglekar became president of the spin-off. Drs. Reklaitis and Okos remained at Purdue as professors and were hired as consultants to the company. Additional research funding was provided by the Indiana Corporation for Science and Technology. These funds have been used to develop a graphics and data-base support system for BATCHES called BPSS. The business has grown from one employee to five, three of whom have doctorates in chemical engineering.

Case Example #4

Applied Physics, Incorporated (API) was founded by Brandon Crowe and David Doty. Crowe was a senior engineer at Purdue's computing center. One of his responsibilities was to repair broken personal computers. When software diagnostics were not possible, troubleshooting the problem could be a very time-consuming process. There did exist at that time a device that test-monitored the activity on the microcomputer circuit bus; however, Crowe was unable to locate such a bus monitor, so he created the "Crowcard." The inventor disclosed his activity and technology to Purdue's Office of Patents and Copyrights. The Patent Office brought Crowe over to INventure because he was interested in starting a business based on the Crowcard. An agreement was finalized with INventure, and API moved into the incubator. At the same time, API signed a licensing agreement with the Office of Patents and Copyrights. Crowe remained at Purdue's computer center for several months before leaving to join API full-time as president.

Purdue's relationship with API represents the ideal situation for creating corporate spin-offs. Purdue receives short-term benefits in the form of a royalty through the license arrangement with API, and long-term benefits through INventure's equity position.

Purdue's licensing agreements are adapted to satisfy the needs of a particular start-up business. API pays no licensing fees for the first two years the product is sold, and in the third year begins paying a nominal percentage. The percentage increases in the fourth and fifth years, before leveling off in year six. Purdue feels its interests are best met if API is allowed to conserve and reinvest its cash in the business. Sacrificing short-term royalties will result in more years of royalties and greater appreciation in API's stock value.

Conclusion

Purdue University's philosophy regarding corporate spin-offs from the university has changed dramatically since the early 1980s. The current policy of encouraging and supporting entrepreneurial activity resulted in the formation of INventure. The incubator assists faculty members who want to commercialize their technologies but who have little or no business experience.

With its change in philosophy has come an easier disclosure and licensing process at Purdue. The university's Office of Patents and Copyrights is now more aggressive in negotiating license arrangements with large businesses, but is also supportive of spin-off companies. The four case examples above demonstrate the basic characteristics of starting and growing a business. All four of the technically oriented founders had to learn the marketing, manufacturing, financial, and administrative aspects of their companies. For the latter two cases, the supportive environment and assistance provided by Purdue University greatly facilitated the faculty entrepreneuring. Such spin-offs are viewed as a win-win situation for the university, faculty, students, and state.

14

Conclusion

Alistair M. Brett, David V. Gibson, and Raymond W. Smilor

The chapters in this volume cover a range of issues, but one theme is common to them all: the search for effective mechanisms for launching and sustaining spin-off ventures. The need may be driven by a desire to form bridging structures to business and industry as described by Cantlon and Koenig, recognizing the university's economic development responsibility. Or the motivating force may be a desire to see direct commercial benefits from university R&D. Gibson and Smilor take this even further in investigating the role of the university in sustaining regional growth, the creation of technopoleis, including new growth companies, and the two-way benefits desired therefrom.

The Italian experience of how a university planted the seed of the Bari technopolis is recounted by Bozzo, Gibson, Sabatelli, and Smilor. An enhanced environment for spin-off ventures was thus created through university, small company, and large company cooperation.

Allen and Norling help allay the fear held by many that involvement in commercial ventures will deflect universities from their traditional mission. Indeed, a second theme that may be identified in this book is how entrepreneurship and faculty commercial endeavors enhance the academic environment. McQueen and Wallmark's chapter reviews some benefits to the university and demonstrates that advantages can greatly outweigh the threats.

Any research commercialization mechanism must contribute to a supportive environment for spin-off company formation. Thus, as noted by Morrison and Wetzel, intellectual property policies—and especially overall faculty support and reward structures—need to be critically reviewed at most institutions.

Much of the above discussion is relevant to the general issue of research commercialization. Factors affecting the choice of transferring technology for maximum benefit through the somewhat traditional licensing route or the more radical spin-off company path are analyzed by Gregory and Sheahen. University and state policies—and especially those designed to help faculty avoid conflict of interest situations—may also be critical in the decision to license or spin-off research results as Wilson and Syzgenda note. Once an effective system is in place for making a choice and allowing a new venture to take shape, then support structures, quasi-independent organizations, external assistance organizations, or all are required to sustain the embryonic venture. Kowalski looks at the basic elements of support and describes a few not-so-obvious pitfalls.

Some principles of breeder formation are provided by Wilem, along with an example of an organization whose aim it is to provide a systematic approach to innovation. The results of successful commercialization—in terms of jobs created and other benefits from specific mechanisms—are discussed by Giannisis, Willis, and Maher in the context of their Illinois program. The topic of university technology commercialization structures is again picked up by Krisst, using the case of the University of Connecticut. Spin-off companies from Purdue University and a university-based support structure are illustrated in Thompson's chapter, demonstrating another effective mechanism.

The University of Texas at Austin and Virginia Tech, among others, are starting to develop detailed case studies of university spin-off companies. There has been a lack of information on both the successes and failures of university spin-offs. Through the current studies, we hope to learn what techniques may be replicated and, perhaps, what problems can be avoided. Unsuccessful spin-off ventures can be just as instructive as successful ones. The research of David Birch and others has shown that many profitable companies have risen out of what was at first apparently a failure. We expect these and future studies to be published as guides for those seeking to learn from the experience of others.

Public and private universities face different opportunities and problems. Existing state conflict-of-interest statutes may constrain the public institution. Most of these regulations, which usually proscribe faculty and staff equity positions in for-profit businesses, were established for legitimate reasons; but they may now have unfortunate consequences for the spin-off venture. The university may wish a faculty member to have a substantial ownership position in a spin-off business—for instance—but also wish to keep the professor at the university. Then too, concern over issues of product liability suits being brought against spin-off ventures—and possible vulnerability of the originating university—are surfacing as commercialization via the spin-off route becomes more common. Risk management is a new issue for many universities. Others question the response of industry. Will corporations become reluctant to sponsor research at an academic institution that is engaged widely in commercial ventures?

Thus, the development of detailed case studies—together with additional research into technology commercialization mechanisms, and clarification of legal issues—is critical before the vast base of university R&D can be beneficially exploited to create wealth, produce regional economic growth and increase national competitiveness. Finally, how might spin-off companies support a nation's international competitiveness? Although such companies are small, many have high growth potential and are based on critical technologies. Strategic alliances with spin-off company partners in other countries may form one way for such ventures to contribute to the global economy.

Appendix A

Venture Capital and the University

M. Campbell Cawood

Business and academia have been perceived as two inherently different and often conflicting concepts.[1] The foundation for the argument is not new. While businessmen have been at odds with those seeking academic freedom for centuries, changes are occurring every day. Some in the academic community now feel left behind by those who have capitalized on research efforts funded by universities. An exciting dialogue is now under way—discussing how academia and business can work together; laying a foundation for changes in what have previously been perceived as conflicting paths.

There are a few trends affecting universities today that accentuate the need for alternative and creative financing. First, the demographics are working against the university. The Baby Boom is over. Universities are now faced with trends of declining enrollment and expanding costs. Traditional budgetary pressures will be even greater in the future since the revenue stream from tuition is declining.

Second, the federal government's role in education and research is diminishing. All types of government spending are being evaluated and reevaluated. Even defense and transfer payments are not truly sacred cows. Education was the first to feel the impact of budget cuts. As long as there are unconscionable deficits and our economic policies flip-flop between deficit spending and inflation, universities should not count on grant money as sustainable. There is a strong and emerging lobby of people who do not believe that university-funded research should be a national priority.

Third, services are moving out of the public sector and into the private sector. Historically the government has been the primary delivery vehicle for services to our society. The public sector delivered the goods. Roles are changing dramatically these days, however. Now, the private service sector is the fastest growing part of our economy, and universities are directly affected. Examples are hospitals and laboratories, many of which are shifting into the private sector. Several of the best biotechnology labs in the country now have "Inc." after their names and are recognized for their academic excellence. Nobel prizes are no longer the sole domain of the university community. In spite of this transition, universities still have a competitive advantage, and a high potential for breakthroughs exists in areas interesting to the business community. It is in the universities that the spark of genius is nurtured.

Fourth, the technological and information society is causing all of us to reevaluate priorities. Technological changes are occurring so rapidly that one cannot afford to wait for a long-term payoff on an investment. Dollars spent today on technology are likely to be obsolete tomorrow, but the cost of deferring those investment decisions is even greater. As a society we are having to learn new skills and thought processes. A large part of this challenge rests with the educational community. Naturally, this means more operating and investment capital is sorely needed to meet the challenge.

Up to this point, we have been discussing challenges. We talked about demographics, the federal government, and how the economy of our country is changing, and asserted that technology will play an even more important role in our future than in the past. Let us now turn to solutions and examine further trends of a more positive nature.

Public Law 96-517

Because of Public Law 96-517, now—for the first time—universities know who owns the commercial rights to grant money funded by the U.S. government. As a result, one of the potential legal obstacles has been removed when a faculty member or a university considers a product or disclosure for commercialization. Now that the government has transferred substantial value to the university, it is the university's challenge to equitably slice the pie between the professors (the workers), and the universities (the owners of the capital). Labor's cut in a mature industry is approximately 40 percent of the revenues and is even more in labor-intensive industries. When we look at service industries—and especially software—labor is the lion's share of the revenue stream.

These observations suggest that the slice of the pie going to those who create commercial products should not be an absolute. Universities need to be sure that proper incentives are structured so talented individuals will not have any excuse to try and find a way around the system—with everyone losing, in the process. All participants in a university-based entrepreneurial venture need to be as clear as possible about contractual agreements with professors—and especially those who are working in science—while allowing room for flexibility. Nothing is advanced when greed takes control. Some professors agree that certain scientific procedures are an art form—unique to them—and cannot be duplicated or patented. Others suggest that the patent office of their university is where you go if you want to be exploited. What this means to me is that they are not going to transfer their knowledge to either the university or a venture capital organization. Policies should facilitate—not deter—creativity. When people know that they are being treated fairly, then the potential for an unlimited return—a good definition of equity—is enhanced.

The Venture Capital Industry

The venture capital industry is very small compared to the size of the publicly traded debt and stock markets. There are, however, now billions of dollars available for investment in private transactions where the investor will be giving up liquidity in exchange for substantial controls over the manner in which the money is spent. When a portfolio manager buys GM stock—for instance—his or her vote hardly counts. But when venture capitalists buy "XYZ" stock, their vote is the difference.

In 1983, $3.4 billion was raised for future commitments, and almost that much was invested. IBM alone spent more than $3 billion in 1983 on research and development. In aggregate, the new business development budgets for large domestic corporations are multiples of these amounts. The difference between what venture capitalists do and what corporations do is in the reward system and in the spirit of entrepreneurism that the former are vigorously developing in the United States.

We also believe that the aggregate numbers for venture investing have not increased so much as has the fact that most of it is now professionally managed. Venture capital has advanced to a state that has left behind the high-flying "try it to see if it works" mentality. It is acquiring new measures of discipline in the process. Indicative of this is that pension funds are now a very large participant in the industry. Basic to the concept is a reduction—not an increase—in risk. When venture capitalists are involved, you are dealing with people who are dedicated to investing every dollar responsibly. Experienced venture capitalists bring substantially more than money to the table.

Several universities have quite successfully invested in venture capital via the limited partnership route. However, it is a very high-risk endeavor for a university to become a venture capitalist. The characteristics of venture capitalism are not generally found within the structured university environment. The following are specific criteria that Venture First Associates in Winston-Salem, North Carolina, uses when considering the funding of a new business.

1. Management. This is the most important criteria. Venture First asks: Does the individual or group of individuals have that unique ability to be an entrepreneur—to inspire leadership and take on what may at times appear to be an impossible task? Or do they measure success by the frequency and amount of their paychecks? Are they willing to make sacrifices for a long period of time to see the business through its development? Many who come to Venture First with ideas and concepts are not capable of running a business. Part of the process is to educate the clients on the elements of a good business and, if necessary, convince them that an experienced manager may be the key to success. It is not Venture First's goal to manage businesses, but rather to see that a capable management team is in place.

2. Product/Services. Exactly what product or service is being proposed? Successful venture capital firms do the proper due diligence to find out if the representations are what the entrepreneur says they are. What is the competition? Does the product really work, or are more development or regulatory processes necessary? Have potential end-users tried it? Can it be sold in the marketplace for a reasonable margin? What are the price and cost sensitivities? Will it fit into a fast-growing market and enjoy a reasonably long life expectancy? Is the fact that it is patented the only defensible part of the business plan?

3. Market. It is important to take a good hard look at the big picture. In what industry or segment of an industry is the focus? Or is a complete new industry being proposed? Is this industry one in which you really want to make a long-term commitment? Look at trends. Can the product fit into an industry in such a way that it will become dominant or continue to gain a market share over the long term? Are the margins such that you can be wrong about the trends and still obtain extraordinary returns? Can a business be structured with the potential of being a $25–100 million business in five to seven years?

4. Finance. How much money is required, and how long will this business take to become profitable on an operational basis? What form will the money take on the balance sheet? Exactly what will the money be used for? (Cashing out previous investors is not the intended purpose of venture capital.) Who are the present investors? How interested (and capable) are they to see the development through the various stages of growth? What is the valuation? Is it in line with comparable marketable valuations and adequately adjusted for risk and return?

5. Influence. Does the entrepreneur share the venture capitalist's goals and standards of doing business? Does the entrepreneur(s) realize that venture capitalists must see vehicles for liquidity in a relatively short time horizon? Will the entrepreneur(s) listen to advice and actively seek it? Will the investors have the ability to replace management in a time of crisis?

These are all areas that venture capitalists scrutinize carefully before investing money. However, if everything went according to the ideal plan, most venture capitalists would not be in business. Now consider what the real world of venture capital is all about:

- Hardly anything works as the inventor expects.
- Inventors usually have no idea who will buy their product/idea or what someone might pay for it.
- Costs of manufacturing or operating controls are foreign concepts to most inventors.
- Most researchers have probably never been in a small business, let alone started one.
- The inventor's family may have lent him or her some working capital;

and family members are still holding on to some form of employment, with commensurate responsibilities.

• Researchers think everyone is trying to cheat them or steal their ideas.

• Control is everything to inventors, and they seldom understand that several rounds of dilution in ownership are necessary when a company is successful and on a fast growth path. It is this growing process that creates the potential for substantial returns.

• Inventors initiate the discussion with venture capitalists using phrases like "loan" and "rate of interest"—thinking of bankers.

• Inventors want enough money for two years and for venture capitalists to leave them alone so they can make the "thing" work.

• Researchers expect venture capitalists to pay for the services they deliver to the new business of which they are part owner at a rate at least equal to their previous salary.

After Venture First wrestles with some of these basic misconceptions, we can then get into the deal itself and call on our experience and imagination. Given our criteria, what is needed to fill in the holes? Do we have the resources—not just financial—to truly add value to the concept and put the puzzle together? It is in this role where seed venture capitalists actually become part of the founding fathers and orchestrate the beginning of a business.

Patent administrators and the university must realize that, if they want a higher return, the trade-off comes in the form of little or no money up front from venture capitalists. Large companies will pay royalties to researchers for licensing agreements, but will seldom share part of the profits. Should a university believe that the market for a certain disclosure is substantial and that a stand-alone business can be built around it, then the university should talk to venture capitalists. The returns can be extraordinary.

Note

Delivered in Dallas, Texas, on September 6, 1984, to the National Council of University Research Administrators at a conference titled "The Private Sector/University Technology Alliance—Making It Work."

Appendix B

Outline for Understanding the Legal Framework of Spin-off Companies

Thomas F. Dowd, Bruce R. Hopkins, and David I. Wilson

In simplistic terms, the process by which technology evolves into a commercially viable business can be divided into three phases. In the first phase, an idea emerges from research in some fashion. At this point, steps should be—and usually are—taken to legally protect the intellectual property inherent in the idea. Potential commercial applications are identified, and often the intellectual property protection is modified to encompass and protect the commercial opportunities. However, the technology has not yet been developed to the point that it would support an operating business and thus induce outside investors to risk capital.

The technology must proceed through the second phase—incubation—during which it is developed through testing, building of prototypes, and other such steps to establish its commercial feasibility. Funding for this process may derive from several sources; and decisions must be made as to the best way to proceed, from a business point of view. Depending on the circumstances, it may make sense to license the idea to a third party for development and wait for an economic return in the form of royalties. It may make sense to try to raise seed capital from sophisticated investors willing to accept a high level of risk in order to maintain the present ownership of the technology. State-sponsored and other institutional incubators are being established that subsidize development in the hopes of creating new industries and job opportunities. Grants are sometimes used. However, large-scale investment by traditional venture capitalists or by public investors is usually extremely difficult to obtain.

The third and final phase is to evaluate the technology as it emerges from the incubator stage, formulate a business that will exploit it, raise the capital necessary to initiate the business, and commence operations.

The boundaries between these phases are not bright lines, and activities germane to each are frequently carried on concurrently. The discussion of legal considerations that follows has applicability to each phase.

Intellectual Property Protection: Obtaining It and Maintaining It

A. Copyright

1. Scope of copyright
   (e.g., of computer software and manuals:
   a. Covers the expression of a computer operating system utilizing a new computer language; or
   b. Does not encompass the idea of a computer operating system for the new language.)

2. Necessary steps for protection
   a. Copyright exists from the moment of the creation of the work.
   b. Notice of copyright should be placed on the work so that it does not pass into the public domain. Publication of the work, such as by distributing it, without the notice can cause the work to become unprotectable.

3. Registration of the copyright with the U.S. Copyright Office—which is inexpensive and can be accomplished relatively quickly—provides the following procedural advantages:
   a. Litigation to prevent infringement of the copyright cannot be undertaken unless there is copyright registration.
   b. The copyright registration constitutes prima facie evidence of ownership of the copyright.
   c. Statutory damages for copyright infringement and attorney fees are available if the work has been registered within three months of the first publication of the work.

B. Semiconductor Chip Designs

Protected by the Semiconductor Chip Protection Act of 1984 in a manner similar to copyrighted work. (The maskworks from which the design is created is the system for protecting chip designs.)

1. Scope of coverage—Chip designs are protected so long as the design is original. Protection does not exist for designs that are staple, commonplace, or familiar, or consist of a variation of a design combined in a way that—considered as a whole—is not original.

2. Necessary steps for protection
   a. Registration with the Copyright Office—which is similar to registering copyrights—must be done within two years.
   b. Use of the proper notice (i.e., the word maskwork; the symbol,*M*; or the letter, M, in a circle) and the name of the owner of the maskwork must accompany the design in circulation.

C. Patents

1. Scope of patents—covers new and useful, nonobvious inventions (e.g., products, processes, apparatuses, and genetically altered multicellular organisms).

2. Patents may be obtained through an examination procedure at the U.S. Patent and Trademark Office. Unlike copyright and maskwork registration, obtaining patents is a complex process.

3. The opportunity to obtain patent protection in the United States can be lost if the invention is described in a printed publication anywhere in the world, or was in public use or on sale in the United States more than one year before the date of the U.S. patent application. Typically, universities agree to a delay in publishing results to permit time to file a patent application (e.g., Carnegie-Mellon reportedly permits a three-month delay if a company insists).

4. A determination must be made as to which inventions have sufficient commercial significance to merit undergoing the patent prosecution process in the United States and in foreign countries.

Consideration should be given to the strategic question of whether or not to seek patent protection because the disclosure required would result in others developing inventions not covered by the patent. In some circumstances, trade secret protection may be a feasible approach to protection, and may be the favored method if disclosure is a problem for some reason.

D. Trade Secrets

Protects knowledge such as a formula, pattern, device, or compilation of information used in one's business that gives the trade secret owner an advantage over competitors who do not know or use it.

1. Scope of trade secrets—encompasses nonpublic information. Trade secret protection is available for copyrighted works, but is not available for patented products or processes. Thus, there is a choice as to whether intellectual property will be protected by the patent or the trade secret laws.

2. Maintenance of trade secret protection requires taking reasonable steps to ensure that secrecy continues.
   a. There must be a preventing of public access to the areas in which the trade secrets can be seen.
   b. The trade secrets should be kept in a secure place.
   c. Employees, advisors, consultants, and so forth should sign agreements that they will not disclose or use the trade secrets except as authorized by the company.
   d. Publications should not disclose trade secrets.
   e. Third parties should be granted access to the trade secrets only on the basis of a written agreement in which the recipient agrees to maintain the secrecy of the information and limits the use of the information.

E. Employee Confidential Information and Inventions Agreement

1. Confidential information
   a. An important element of the protection of trade secrets is an agreement between the company and each employee that the employee will maintain the confidentiality of and not use the company's secrets both during and after his or her employment.

   b. The contract should regulate the information that employees take with them when they leave. (For example, postdoctoral students have taken proprietary cell lines with them so that they can continue their research.)

   c. Unclear agreements lead to litigation. (For example, Caltech had litigation concerning whether a computer program language was owned by the university or by the professor. Caltech's policy was that patentable properties belong to it, but copyrightable properties belong to the professor. The issue was whether the computer language should have been patented, or whether it should have been copyrighted.)

2. Inventions disclosure and ownership agreement

   a. The employee must disclose to the employer inventions, ideas, and processes that result from the work performed by the employee and involving the use of the employer's facilities, inventions, or confidential information.

   b. If the employee may obtain rights in the inventions, the rights should be clearly stated.

F. Rights of Nonprofit Organizations and Small Business Firms to License Patents Funded by the Federal Government

1. Nonprofit organizations and small business firms must consider the impact of federal statutes and policies with regard to obtaining title to patentable inventions that are conceived or reduced to practice during the performance of work under a federal research contract. They need to distinguish between inventions funded by the federal government and those inventions that are not covered, by the requirements discussed below. For example, the government would have no rights where an instrument purchased with government funds is later used—without interference with or cost to the government-funded project—in research or development all expenses of which involve nongovernment funds.

2. Upon disclosure of an invention and application to the federal agency funding the research, nonprofit organizations and small business firms can receive title to their patentable inventions. The government reserves both a nonexclusive, nontransferable, irrevocable, paid-up license to practice the invention and the right to exercise "march-in" rights to grant licenses under the patent if

steps are not taken to achieve practical application of the invention, or if health or safety needs are not reasonably satisfied by the firm or their licensees.

3. The statute governing government-funded patents places restrictions on transfer of patent rights. The act mandates a preference for having the product manufactured substantially in the United States. Also, the patent may not be assigned without approval from the federal agency. While exclusive licenses may be granted to small business firms, others may receive a grant of exclusivity for five to eight years.

Licensing Intellectual Property

A. Business Reasons for Licensing

1. The business objectives of the license agreement—which should be the subject of a business plan—will govern the approach taken to the terms of the license. Before drafting a formal agreement, the key elements of the agreement should be negotiated and reduced to writing in a term sheet of some sort.

2. Some of the classic reasons for licensing
   a. To obtain royalties for uses in which the licensor is not interested, but that may be valuable to third parties (e.g., Xerox licensed xerographic patents for use in X-ray equipment to General Electric)
   b. To obtain royalties for products in which the licensor is interested but lacks the resources to exploit (e.g., license intellectual property rights in foreign countries)
   c. To obtain cross-licenses to intellectual property rights from other companies (e.g., another company owns rights in blocking patents—patents necessary for the licensor to produce a workable product)
   d. To generate revenues to fund operations, development, and so on.
   e. To gain faster entry into the market or the use of another company's expertise to obtain such entry or reduce risks (e.g., license intellectual property covering a medical device or product to a company experienced in obtaining regulatory approval from the FDA)

B. The License Agreement Issues

1. Subject matter of the license agreement
   a. The importance of clear definitions of technical terms
   b. The need to specify the intellectual property licensed (e.g., patents or patent applications, or trade-secret data package)
   c. Field of use (e.g., a pharmaceutical licensed to one company for use in humans and to another for use in animals)

2. Geographic scope of the license

3. Option rights (e.g., six-month option to consider entering into a license; option to renew for a further period at a higher payment)

4. Rights granted may include one of more of the following:
   a. Right to make, use, or sell for the life of the patent, or for a perpetual period in the case of a trade-secret data package
   b. Right to make, use, or sell for a limited period of time
   c. Exclusive versus nonexclusive license
   d. Right to assign or sublicense
   e. Future developments

5. Licensee obligations may include one or more of the following:
   a. Up-front payment
   b. Minimum royalties or specified revenue goals
   c. Royalties that increase or decrease depending on the number of units sold or the passage of time; designation of the base used to compute (e.g., net sales)
   d. A percentage of the licensee's equity, or percentage of the licensee's income
   e. Nonexclusive license to use of patents by the university company, but a sharing of royalties earned from third companies
   f. Subject to licensor march-in rights (i.e., the right to reclaim from the licensee all or certain rights in the licensed intellectual property rights—or to require mandatory sublicensing by the licensee—if the licensee does not take specified steps to develop and market the licensed invention in a timely manner)
   g. Consideration of adjustment of royalty rates depending on the nonissuance of patents or invalidity of licensed patents
   h. Requirement to market in specified manner (e.g., exhibit software at listed trade shows)

    i. Performance of necessary testing protocols (e.g., for FDA approval)

    j. Obtaining of necessary regulatory approvals

    k. Requirements that the licensee purchase products from the licensor or sell products to the licensor

6. Grant backs of licenses to improvements

7. Responsibility for filing and prosecuting patent applications based on the licensee's inventions (e.g., if the licensee does not want to file an application, the licensor is granted the option to do so)

8. Royalty reports and right of audit

9. Protection of intellectual property rights (see "Intellectual Property Protection" above)

10. Warranties or disclaimers of warranties on performance and warranties regarding noninfringement

11. Obligations regarding suing unlicensed infringers

12. Technical assistance (e.g., licensor will make available scientific assistance of a specified level of expertise for stated periods of time at fixed daily compensation rates)

13. Software issue: Should the source code of the licensor be disclosed to the licensee? Should the source code be placed with an escrow agent?

14. Termination of the license
    a. Life of the patents; may need to reduce the royalty rates if patents expire at a staggered rate
    b. Shorter terms for exclusive licenses
    c. Failure to meet specified sales or performance goals
    d. Bankruptcy or other forms of insolvency
    e. Need for return of trade secret information upon termination of the agreement

15. Procedures for resolution of disputes through arbitration

Basic Organizational Considerations in the University Setting

A. Structural Matters

1. Nature of high-technology activity

2. Determining form of entity affiliated with the university
   a. Operating internally
   b. Use of nonprofit/for-profit subsidiary
   c. Joint venture/partnership

3. Other management elements
   a. Capitalization
   b. Governments structure
   c. Ongoing revenue flow

B. Tax Considerations in the University Setting

1. Tax exemption rules (see Appendix C, which follows)

2. Present unrelated income rules
   a. Trade or business test
   b. Regularly carried on test
   c. Substantially related test
   d. Activity exceptions
   e. Income exceptions (generally for passive income)

3. Coming unrelated business rules
   a. Attribution of subsidiaries activities
   b. Joint ventures
   c. Research exception
   d. Royalties rule

Formulation of the Spin-off Business

In order to transform technology in the incubator phase to an operating business, it is necessary to develop a relatively detailed written operational and financial plan. This document is usually referred to as a business plan, but it may also be called an investment or funding proposal, an offering memorandum, an investment presentation, or some other similar title.

The purposes of the business plan are to make the entrepreneur think about:

1. The manner in which the business is to be operated;

2. The funds that will be required to commence and prosecute those operations;

3. The sources of funds and the other resources that are available;

4. The market in which the business will operate;

5. The management team and other personnel necessary to carry out the business; and

6. The profit potential of the business.

Investors will not seriously consider a proposal until a plan of this type has been prepared and often will not even meet to consider the investment until after a review of the plan. The plan should be comprehensive, but not so exhaustive that it is overwhelming. The textual portion and financial statements and projections should total approximately 50 pages. The plan should contain the following information

A. Brief Summary Discussion of All the Plan's Elements

1. The business—the products and services offered

2. The company and its management

3. The market—size and type of customers

4. Principal strategies

5. Primary competitive considerations

6. Primary investment risks

7. Financial summary and highlights

B. Detailed Description of the Company

1. Business concept—services and products

2. Operating history and current status

3. Unique skills, technologies, and other resources

C. Market Position and Evaluation

1. Customers

2. Market size and trends

3. Competition

4. Existing market share and projected penetration

5. Ongoing monitoring and evaluation of market
   a. Trends
   b. New entrants
   c. New technologies

D. General Strategies

1. Statement of goals

2. General strategies to accomplish goals

3. Marketing strategy
   a. Categories of customers
   b. Advertising and promotion
   c. Sales development

4. Competitive strategies

5. Strategic alliances

6. Operational strategies
   a. Pricing
   b. Geographical location and service areas

c. Facilities and improvements
d. Personnel requirements

E. Organization and Management

1. Present capital structure and ownership interests

2. Executive officers and other key management personnel

3. Other key personnel (e.g., research scientist)

F. Other Investment Considerations

1. Risks

2. Regulatory matters (if any)

G. Financial Information

1. Historical (if available)
    a. Balance sheets
    b. Operating statements—income and cash flow
    c. Statement of working capital requirements

2. Projections
    a. Major assumptions
    b. Pro forma income statements
    c. Pro forma balance sheets
    d. Pro forma cash-flow statements (showing sources and uses of funds)
    e. Debt service requirements

H. Structure of Proposed Financing (if appropriate)

1. Valuation of enterprise

2. Type of securities to be purchased; percentage of equity to be sold

3. Timing of investment

Structuring the Spin-off Enterprise

A. Investment Structure

1. Type of security purchased
   a. Debt
   b. Equity—common stock or preferred stock
   c. Convertible debt or stock
   d. Warrants, options
   e. A "Strip" of different types of securities

2. Timing of investment
   a. Coordination with funding needs, to minimize equity sold
   b. Investor control of amount of capital drawn

3. Valuation methods
   a. Projections
   b. Rate of return/risk analysis
   c. Cash flow; terminal value; value of assets

4. Consideration for investment
   a. Cash
   b. Property; valuation
      i. Tangible
      ii. Intangible
   c. Sweat equity—past and future
   d. Other future services
      i. Operational
      ii. Advisory; consultive
      iii. Financial
      iv. Vendor or customer

5. Tax matters
   a. Section 351 issues—property versus services; control
   b. Debt/equity issues
   c. Start-up costs
   d. Intangible and tangible assets
      i. Acquired for cash, notes, and so on
      ii. Acquired for equity
   e. Partnership (general or limited) versus corporation; Subchapter S issues
   f. Original issue discount; imputed interest

6. Sources of funds
   a. Founders and friends
   b. Investors
   c. Institutional lenders
      i. Banks, thrifts
      ii. Insurance companies
      iii. Pension funds
      iv. Other institutions
   d. Vendor's payment terms
   e. Customers (e.g., prepayments, deposits)
   f. Leases; installment sales
   g. Federal and state agencies or instrumentalities
      i. Direct funding
      ii. Credit enhancement
      iii. Rate improvement

7. Dilution issues
   a. Future draws
   b. Future financings
   c. "Earn-out" provisions

B. Allocation of Control

1. Allocation of functions, and control thereof, among two or more entries (e.g., manufacturing and sales)
   a. Board of directors
      i. Number, composition
      ii. Mechanics of designation
         (1) Class vote
         (2) Stockholder agreement; proxies
         (3) Voting trust
      iii. Mechanics of removal
      iv. Cumulative voting
   b. General partner (managing GP)
      i. Corporate
      ii. Mechanics of designation
      iii. Mechanics of removal
      iv. Percentage versus per capita voting
   c. Contractual allocation (e.g., close corporation agreement)

2. Decisions concerning major structural matters
   a. Mergers, acquisitions

    b. Divestitures; sale of assets; sale of divisions or subsidiaries
    c. Redemptions or stock repurchases
    d. Recapitalizations
    e. Revisions of fundamental government documents
    f. Sale of business
    g. Future sales of equity or debt to additional investors
        i. Private transactions
        ii. Going public

3. Decisions concerning major operational and financial matters
    a. Hiring and firing of key personnel
    b. Debt/equity ratio
    c. Budget approval
    d. Accounting and reporting systems
    e. Lines of business—additions or deletions
    f. Capital expenditures
    g. Transactions with affiliates
    h. Employee incentive plans
    i. Research and development

4. Dispute resolution
    a. Arbitration
    b. "Push/pull" provision
    c. Reciprocal puts and calls at designated price or formula, or using specified procedures
    d. Dissolution
    e. Right of first refusal; right to match

5. Control mechanics
    a. Define scope and purpose of business in governing documents
    b. Veto or supermajority voting provisions
    c. Quorum requirements
    d. Rights to participate
        i. First refusal
        ii. Cosale
        iii. Preemptive rights
    e. Classified or unequal voting ("A–B capitalization")
    f. Specific provisions in governing documents or contract

C. Allocation of Economic Benefits

1. Distributions to investors
    a. Dividends
    b. Interest and principal—debt (straight or convertible)

    c. Return of capital
    d. Timing and control of payouts

  2. Compensation (cash or stock)
    a. Management or consulting fees
    b. Royalties
    c. Salaries, bonuses
    d. Fees for other services
    e. Commissions

  3. Computation and priorities of payouts
    a. Percentage allocation of profits and losses
    b. Priority in payments
       i. Contractual payments
      ii. Debt
     iii. Equity
    c. Flips in profit-sharing percentages

  4. Treatment of expenses
    a. Past—payment versus credit toward investment
    b. Future—payment versus credit toward investment
    c. Organizational
    d. Operational subsidies or favorable terms

  5. Liquidity
    a. Restrictions on transfer of shares
    b. Reciprocal puts and calls
    c. Going public
    d. Sale of business

D. Personnel Matters

  1. Employment agreements

  2. Remuneration
    a. Salaries
    b. Cash incentives
    c. Deferred compensation plans
    d. Perquisites
       i. Individual
      ii. Group plans
    e. Tax and ERISA issues

3. Stock incentives
   a. Stock purchase plans
   b. Stock bonus plans
   c. Stock option plans
   d. Stock appreciation rights; "phantom" stock
   e. Tax matters
      i. Deferral
      ii. Section 83 matters
      iii. Qualified versus nonqualified
   f. Securities law issues

4. Retirement planning
   a. Pension—defined benefit versus defined contribution
   b. Profit sharing
   c. Supplemental benefits
   d. Tax and ERISA issues

5. Protective measures
   a. Agreements not to compete
   b. No solicitation of employees or customers
   c. Trade secret and intellectual property matters
   d. "Golden parachutes"
      i. Terms
      ii. Tax analysis
   e. Vesting schedules for stock incentives and other benefits
      i. Time limits
      ii. Performance tests

E. Other Legal and Accounting Considerations

1. Federal and state securities law compliance regarding offer and sale of investment
   a. Registration of securities; exemptions
   b. Broker/dealer issues
   c. Investment advisors
   d. Investment company
   e. Restrictions on transfer; registration rights
   f. Disclosure requirements

2. Fiduciary duties—partnerships versus corporations

3. Accounting matters
    a. Pooling versus purchase
    b. Debt/equity
    c. Treatment of start-up expenses
        i. Organizational costs
        ii. R&D, patents, trademarks, and so on
    d. Other developmental-stage issues
    e. Financial reporting generally
        i. Timing issues
        ii. Consolidation; equity accounting
        iii. Selection of principles
        iv. Audited v. unaudited

4. Antitrust
    a. Exclusive territories; other exclusive dealings
    b. Price maintenance
    c. Noncompetition agreements
    d. Hart–Scott–Rodino
        i. Control issues; ultimate parent
        ii. Definition of market

5. Due diligence matters
    a. Legal review
    b. Accounting review
    c. Business review
    d. Corrective measures

6. Creditor's rights analysis
    a. Fraudulent conveyance issues
    b. Liens and so forth

7. Regulatory matters—federal and state
    a. Type of industry; specific product or service
    b. Licensing
    c. Clearances and consents
    d. Timing considerations

8. Liability/litigation analysis
    a. Areas of exposure
    b. Insurance coverage; general liability

    c. Effect of type of entity
       i. General partnership or joint venture
      ii. Limited partnership
     iii. Corporation
    d. Contractual relationships
    e. Other protective steps (e.g., disclaimers, releases, separate subsidiary)
    f. D&O insurance

F. Structural Relationship

1. Corporation

2. Limited partnership

3. General partnership

4. Joint venture

5. Licensor/licensee

6. Franchisor/franchisee

G. Typical Documents

1. Corporation
    a. Articles of incorporation; stock terms
    b. By-laws
    c. Preincorporation agreement (optional)
    d. Subscription or stock purchase agreement
    e. Stockholders' agreement
    f. Director and stockholder actions
    g. Close corporation agreement (if appropriate)

2. Partnership
    a. Partnership agreement
    b. Certificate of partnership
    c. Actions by partners

3. Debt documents
   a. Loan agreement
      i. Terms; future advances
      ii. Security
      iii. Representations and warranties
      iv. Restrictive covenants
      v. Events of default
      vi. Equity kicker
   b. Notes, debenture, and so forth
      i. Interest—rate; payment schedule
      ii. Principal—amortization; right to prepay
      iii. Subordination
      iv. Security; guarantees
      v. Remedies
   c. Security documents
      i. Security agreement
      ii. Pledge
      iii. Mortgage; deed of trust
      iv. Financing statements
      v. Special documents (e.g., patents, trademarks, ships, aircraft)
   d. Intercreditor agreement
      i. Subordination
      ii. Right to exercise remedies; lien priorities
   e. Indenture
   f. Participation agreement

4. Operational contracts
   a. Sales contracts
   b. Services contracts
   c. Contracts with vendors (e.g., OEM agreements)
   d. Sales representative and distribution requirements
   e. Licensing agreements

5. Timetable and responsibility checklist

Appendix C

How to Deal with Unrelated Business Income Tax

Deborah Walker

A college or university that satisfies the requirements of Section 501(c)(3) of the Internal Revenue Code is exempt, in general, from federal income taxation. Under this section, as long as the institution engages in activity that relates to its educational and other exempt (such as basic research) purposes, the income from those activities will escape taxation. Many colleges and universities, however, engage in other activities that raise funds to be used in the exempt purpose, but are not themselves related to that purpose. Many of these activities are the type that are normally carried out in the private sector by taxpaying organizations. To eliminate what was perceived as an unfair competitive advantage for tax-exempt organizations, Congress instituted a tax on "unrelated business taxable income" (UBTI) in the Revenue Act of 1950.[1] This act is important to tax-exempt organizations for two reasons: The tax represents an additional expense reducing cash flow to the organization. More importantly, excessive unrelated business income can jeopardize tax-exempt status for the entire organization.

Is It Unrelated Business Taxable Income?

A college or university is deemed to have unrelated business income when it derives income from any regularly conducted trade or business that is not substantially related to its educational and exempt purpose.[2] An analysis to determine whether an activity produces unrelated business taxable income requires resolution of the following questions:

1. Is the activity a trade or business?

2. Is the activity "regularly carried on"?

3. Is the activity "substantially related" to the exempt purpose of the college or university?

Trade or Business

A trade or business in this context is generally defined the same as it is elsewhere in the Internal Revenue Code: Any activity carried on for the production of income from the sale of goods or the performance of services.[3] For testing purposes, each activity is "fragmented"—that is, an integrated group of activities conducted by a college or university is broken down into component activities, and each activity tested separately to determine whether it is an unrelated trade or business.

A college or university that is in a partnership that is engaged in a trade or business is considered to have trade or business income attributable to its share of partnership income. As far as this rule is concerned, it does not matter whether the school is a general or limited partner.[4]

Regularly Carried On

A specific business activity will be deemed to be regularly carried on if it is conducted with a frequency, continuity, and manner of pursuit comparable to the conduct of the same or similar activity by a taxable organization. An activity will be considered *not* to be regularly carried on if it is conducted: (1) on a very infrequent basis (once or twice a year); (2) for only a short period of the year, whereas it would normally be conducted by a taxable organization for a much longer period; or (3) without the competitive and promotional efforts typical of commercial endeavors.

Substantially Related

Income from an activity that is a trade or business that is regularly carried on will not be classified as UBTI if the activity is substantially related to the exempt purpose of the college or university. The activity must have a substantial causal relationship to the achievement of the exempt purpose, other than the mere production of income to support such purpose. In general, only a trade or business operated as an integral part of a university's educational program is considered related. For instance, a university newspaper operated by its students may publish paid advertising. The solicitation, sale, and publication of the advertising are conducted by students, under the supervision and instruction of the university. The regulations indicate that, even though the services rendered to the advertisers are of a commercial nature in this example, the advertising business contributes importantly to the university's educational program through the training of the students involved. Hence, none of the income derived from publication of the newspaper constitutes gross income from unrelated trade or business.[5]

When Does an Organization Jeopardize Its Tax-Exempt Status?

An organization will maintain tax-exempt status under Section 501(c)(3) only if it is "operated exclusively" for one or more exempt purposes.[6] An organization will not be so regarded if more than an unsubstantial part of its activities does not further its stated exempt purpose. The determination of whether engaging in a regular business of a kind ordinarily carried on for profit is merely incidental to the exempt purpose depends on the facts and circumstances in each case. The courts and the IRS have been somewhat imprecise in the standard of measurement to be used in this regard. One possible measurement is the time that employees spend on the activity. Another possible measurement is the receipts and disbursements related to the activity. Both time and financial data were factors in *Associated Master Barbers and Beauticians of America, Incorporated* 69 T.C. 53 (1977), although the court did not discuss how it applied these measurements together.

Several strategies are available to avoid loss of tax-exempt status because of excessive unrelated business income. Certain types of income that are mostly passive in nature are statutorily excluded from unrelated trade or business income classification; this income includes dividends, interest, annuities, royalties, rental income from real property, and capital gains. Whenever possible, tax-exempt organizations should consider transforming active trade or business income into one of these excluded passive types. Permitting the use of property—rather than an outright sale—in exchange for rent or royalty payments will shield the income from UBTI consideration.

Forming a Taxable Subsidiary

One of the ways that tax-exempt organizations can avoid problems with excessive UBTI is the formation of a taxable, for-profit subsidiary to engage in activities that may not be related to their exempt purpose. Provided that certain conditions exist to maintain the separate character of the subsidiary, generally the activities of a taxable subsidiary will not jeopardize the parent's exempt status. In GCM 39598, the IRS held that the leasing activities of an exempt subsidiary in a seven-entity health care system not only resulted in unrelated business income, but were extensive enough to result in private inurement—resulting in revocation of the subsidiary's exempt status. The IRS ruled, however, that the inurement taint would not be attributed to the parent or other affiliates, but would be limited to the subsidiary.

In principle, a parent corporation and its taxable subsidiary will be considered separate entities so long as the purposes for which the subsidiary is incorporated are the equivalent of business activities or the subsidiary subsequently carries on business activities.[7] A corporate subsidiary must be organized with the bona fide intention that it will have some real and substantial business function.[8]

The corporate entity of the subsidiary may be disregarded if the parent exerts so sufficient a control as to render the subsidiary an instrument of the parent. In the case where a subsidiary is formed for a valid business purpose, the activities of the subsidiary cannot be attributed to its parent unless the facts and circumstances provide evidence that the subsidiary is merely an agent for the parent. Situations where the parent is deemed to be controlling the subsidiary would include where the parent is involved in the day-to-day management of the subsidiary, where the subsidiary's board of directors has no independence of action, or where transactions between the two entities are conducted on less than an arm's-length basis.

Income from Taxable Subsidiary—When Is It UBTI?

While formation of a taxable subsidiary may remove the immediate threat to the parents' tax-exempt status, the question of excessive UBTI is not completely resolved. A new issue comes into play concerning the taxability of the income that the exempt parent receives from the taxable subsidiary. The general rule of Section 512(b) excludes most passive income—such as dividends, interest, annuities, royalties, capital gains, rents from real property, and to a limited extent the rents from personal property—from treatment as UBTI. However, in a case where Section 512(b)(13) applies, most items found in Section 512(b)(1) (with the exception of dividends) and all items found in Section 512(b)(2) and (b)(3) are taxable. Under Section 512(b)(13), the exclusion of interest, annuities, royalties, and rents from UBTI status does not apply where such amounts are derived from controlled organizations—whether or not the activities from which such amounts are derived represent a trade or business or are regularly carried on. The income from a "controlled subsidiary" is taxable to the controlling parent at a specified ratio depending on whether it is income from exempt or nonexempt functions.

What is "control" in this regard? The term "control" is defined in Section 368(c) as ownership of at least 80 percent of the total combined voting stock and at least 80 percent of the total number of shares of all other classes of stock in the corporation. Control of a nonstock corporation means at least 80 percent of the directors or trustees of such organization are either representatives of or directly or indirectly controlled by the controlling (parent) organization.[9] If control of the organization is acquired or relinquished during the taxable year, only the interest, annuities, royalties, and rents paid or accrued to the controlling organization for that portion of the year during which it has control are subject to UBTI tax.

Generally, a tax-exempt parent will own 100 percent of a taxable subsidiary—a situation that automatically triggers the provisions of Section 512(b)(13). Apparently, direct ownership is critical; that is, attribution rules are not applicable in this area. In a recent IRS letter ruling (TSM 8439001), the service held that interest payments received by a tax-exempt organization from its second-tier taxable subsidiaries were not subject to Section 512(b)(13) because the subs were not directly owned by the exempt organization. This finding conflicts with the earlier interpretation invoked in GCM 38878 (July 16, 1982), which concluded that Section 512(b)(13) was sufficiently broad in scope to cover a situation where wholly owned subsidiaries transact business among themselves and not directly with the parent. Thus, a mechanical test to determine control under the provisions of Section 512(b) may not always be the final resolution of the issue.

Why Not Dividends?

In essence, Section 512(b)(13) was designed to prevent a subsidiary from disguising otherwise unallowable payments to its parent as tax deductible items. The legislative language in the Tax Reform Act of 1969 points out the then-existing loophole:

In certain cases exempt organizations do not engage in business directly but do so through nominally taxable subsidiary corporations. In many such instances the subsidiary corporations pay interest, rents, or royalties to the exempt parent in sufficient amounts to eliminate their entire income, which interests, rents, and royalties are not taxed to the parent even though they may be derived from an active business.[10]

Since dividends are distributions to shareholders out of a corporation's current or accumulated after-tax earnings and profits, they are generally not subject to intercompany manipulation. The ultimate classification of a payment from subsidiary to parent as a nontaxable dividend or taxable interest falls under the general corporate rules.

Calculation of UBTI Received from a Taxable, Controlled Subsidiary

Interest, annuities, and royalties paid from a controlled taxable subsidiary to an exempt parent are taxed as UBTI according to the percentage of the income as if the controlled subsidiary were exempt. The income is multiplied by a fraction, the numerator of which is "excess taxable income." The denominator is the greater of taxable income of the controlled organization or excess taxable income. Both of these amounts are determined without regard to any amount paid directly or indirectly to the controlling organization. "Excess taxable income" means the excess of the controlled organization's taxable income over the amount of taxable income that, if derived directly by the controlling organization, would not be unrelated business taxable income.[11] Section 512(b)-1(1)(3)(iii) contains a discussion of the following examples.

Example 1

"A," a 501(c)(3) exempt university, owns all the stock of "M," a nonexempt organization. During 1971, M leases a factory and a dormitory from A for a total annual rent of $100,000. During the taxable year, M has $500,000 of taxable income, disregarding the rent paid to A: $150,000 from a dormitory for students of A university, and $350,000 from the operation of a factory that is a business unrelated to A's exempt purpose. A's deductions for 1971 with respect to the leased property are $4,000 for the dormitory and $16,000 for the factory. Under these circumstances, $56,000 of the rent paid by M will be included by A as net rental income in determining its UBTI, computed as follows:

M's taxable income (disregarding rent paid to A)	$500,000
Less taxable income from dormitory	150,000
Excess taxable income	$350,000
Ratio ($350,000:$500,000)	7/10
Total rent paid to A	$100,000
Total deductions ($4,000 + $16,000)	20,000
Rental income treated as gross income from an unrelated trade or business (7/10 of $100,000)	70,000
Less deductions directly connected with such income (7/10 of $20,000)	14,000
Net rental income included by A in computing its UBTI	$56,000

Example 2

Assume the facts as stated in Example 1, except that M's taxable income (disregarding rent paid to A) is $300,000 consisting of $350,000 from the operation of the factory and a $50,000 loss from the operation of the dormitory. Thus, M's excess taxable income is also $300,000, since none of M's taxable income would be excluded from the computation of A's UBTI if received directly by A. The ratio of M's excess taxable income to its taxable income is therefore one ($300,000:$300,000). Thus, all the rent received by A from M ($100,000) and all the deductions directly connected ($20,000) are included in the computation of A's UBTI.

In the case where the controlled nonexempt corporation has a net operating loss without regard to the amounts paid to a controlling exempt organization, the IRS has ruled that Section 512(b)(13) is still applicable.[12] The parent corporation is taxable regardless of whether a current benefit for the corresponding deduction is available to the subsidiary. The IRS concluded that the ratio of excess taxable income to the greater of taxable income or excess taxable income could still be one, resulting in inclusion of the amounts paid or accrued by the parent in its taxable income—even if the subsidiary was in a loss position.[13] In this situation, then, operations cost the parent more than would have been the case had there been no subsidiary.

Section 514: Bad News Gets Worse

Section 512(b)(4) provides that income from debt-financed property is taxable as UBTI, regardless of whether this income is in the form of dividends, interest, rents, or other types of nonbusiness income. "Debt-financed property" is defined as any property held to produce income for which there is "acquisition indebtedness" at any time during the taxable year, or for which there was acquisition indebtedness at any time during the 12-month period ending with the date of disposition if the property was sold during the year.

The term "acquisition indebtedness" means any indebtedness incurred in acquiring or improving the property, incurred before the acquisition or improvement of the property if it would not have been incurred but for such acquisition or improvement, or incurred after the acquisition of the property if such indebtedness would not have been incurred but for such acquisition or improvement and if the incurrence of the indebtedness was reasonably foreseeable at the time of the acquisition or improvement. Section 514's reach is potentially broader than Section 512(b)(13), as it affects dividend income and income described in IRC (512)(b)(5). A computation under Section 514 may produce liability in addition to liability under Section 512(b)(13), in a situation where both subsections apply.[14]

Example 3

"Z," an exempt university, owns all the stock of "M," a nonexempt corporation. During 1971, M leases from Z a university factory unrelated to Z's exempt purpose, and a dormitory for the students of Z, for a total annual rent of $100,000: $80,000 for the factory and $20,000 for the dormitory. During 1971, M has $500,000 of taxable income, disregarding the rent paid to Z: $150,000 from the dormitory and $350,000 from the factory. The factory is subject to a mortgage of $150,000. Its average adjusted basis for 1971 is determined to be $300,000. Z's deductions are $4,000 for the dormitory and $16,000 for the factory. Section 514 applies only to the portion of the rent that is excluded from the computation of UBTI by operation of Section 512(b)(3) and not included in such combination pursuant to Section 512(b)(13). Since all the rent received by Z is derived from real property, Section 512(b)(3) would exclude all such rent from computation of Z's UBTI. However, 70 percent of the rent paid to

Z with respect to the factory and 70 percent of the deductions directly connected with such rent will be taken into account by Z in determining its UBTI pursuant to Section 512(b)(13), computed as follows:

M's taxable income (disregarding rent paid to Z)	$500,000
Less taxable income from dormitory	150,000
Excess taxable income	$350,000
Ratio ($350,000:$500,000)	7/10
Total rent paid to Z	$100,000
Total deductions ($4,000 + $16,000)	20,000
Rental income treated under Section 512(b)(13) as UBTI (7/10 of $100,000)	70,000
Less deductions directly connected with such income (7/10 of $20,000)	14,000
Net rental income included by Z in computing UBTI pursuant to Section 512(b)(13)	$56,000

Since only that portion of the rent derived from the factory, and the deductions directly connected with such rent not taken into account pursuant to Section 512(b)(13), may be included in computing UBTI by operation of Section 514, only $10,000 ($80,000 less $70,000) of rent and $2,000 ($16,000 less $14,000) of deductions are taken into account. The portions of such amounts to be taken into account is determined multiplying the $10,000 of income and $2,000 of the deductions by the debt/basis percentage. The debt/basis percentage—the ratio in which the average acquisition indebtedness ($150,000) is over the average adjusted basis—is 50 percent for 1971. Under these circumstances, Z will include net rental income of $4,000 in UBTI for 1971, computed as follows:

Total rents	$10,000
Deductions directly connected with such rents	2,000

Debt/basis percentage ($150,000/$300,000)	50%
Rental income treated as gross income from unrelated trade or business (50% of $10,000)	5,000
Less the allowable portion of deductions directly connected with such income (50% of $2,000)	1,000
Net rental income included by Z in computing its UBTI under Section 514	$4,000

In a situation where both Section 512(b)(13) and Section 514 apply, Section 512(b)(13) must be considered first, because Section 514(b)(1)(B) makes section 514 inapplicable to income taxed under any other provision of Sections 511 through 514. Section 514 is only applicable if an organization is otherwise not liable for tax, or its tax liability would be increased by the application of Section 514.

Certain assets are not deemed to be debt-financed property, even though there is acquisition indebtedness. These include: (1) any property 85 percent or more of the use of which is substantially related (aside from the need for income) to the performance of the tax-exempt purpose—but if less than 85 percent of its use is substantially related, then only the portion so used will not be treated as debt-financed property—(2) any property used in a trade or business that is not an unrelated trade or business, and in certain research activities; and (3) any property to the extent that income from it is treated as UBTI. The gain on the sale of such property is still subject to the debt-financed property rules.

Summary

Tax-exempt organizations must take care to structure their corporate structure and intercompany dealings to avoid the issues of excessive UBTI and private inurement. Controlled subsidiaries may or may not be a viable means of shifting UBTI away from the parent organization. A parent corporation that can provide operating capital in the form of capital corporations for a new entity and avoid all intercompany debt and cost sharing may skirt the problem altogether. Debt-financed property held by either parent or subsidiary entities can generate additional UBTI.

Dealings between corporate shareholders must be undertaken at arm's length to avoid any taint of private inurement.

Tax-exempt organizations have been the focus of increased scrutiny by Congress, the Treasury Department, the IRS, and many state legislatures. The House Ways and Means Oversight Subcommittee is considering recommendations for changing certain unrelated business income tax provisions. These recommendations have not been finalized, and more hearings are expected. The Treasury Department favors changes regarding the definition of a controlled subsidiary and the aggregation of all of an exempt organization's activities. The Internal Revenue Service favors extension of private-foundation rules to limit the business activities of exempt organizations.

The IRS has increased the intensity and frequency of its examinations of colleges and universities, placing them at a higher level of exposure to substantial tax deficiencies. Therefore, management and trustees should periodically assess the exposure of their institution to unrelated business income tax. A proactive approach would suggest that colleges and universities plan transactions to avoid this tax. As the Internal Revenue Service becomes ever more real to the 501(c)(3) institutions, such proactive planning will undoubtedly be rewarded with lower tax assessments.

Notes

1. H.R. 2319, 81st Cong., 2nd sess. 36 (1950); S.R. 2375 81st Cong., 2nd sess. 28 (1950).
2. IRC Section 512(a)(1), 513(2).
3. IRC Section 162.
4. Reg. Sec. 1.512(c)(1)-1.
5. Reg. Sec. 1.513-1(3)(4)(iv), example 5.
6. Reg. Sec. 1.501(c)(3)-1(c)(1).
7. *Moline Properties, Incorporated* v. *Comm.,* 319 U.S. 436, 438 (1943); *Britt* v. *United States,* 431 F.2d 227, 234 (5th Cir. 1970).
8. Reg. Sec. 1.512(b)-1(1)(4)(i)(b).
9. H.R. Rep. No. 91-413, pt. 1, 91st Cong., 1st sess. 49 (1969), 1969-3 C.B. 232.
10. Reg. Sec. 1.512(b)-1(1)(3)(ii).
11. PLR 8439001.
12. Reg. Sec. 1.514-(b)-1(b)(3), example 3.
13. GCM 39286.
14. Reg. Sec. 1.503-(b)-1(2).

About the Sponsors

Virginia Polytechnic Institute and State University

As a major research university with annual research expenditures of over $100 million and currently ranked 44th among the top 100 such research institutions, Virginia Polytechnic Institute and State University (Virginia Tech) has become a leader in promoting the wide benefits of commercializing university research and development. Some 16 percent of the university's research is directly sponsored by industry—one of the highest proportions in the nation.

Virginia Tech's Office of Technology Management and Transfer is developing case studies of successful research commercialization at U.S. and foreign universities. Studies of transfer and commercialization of technologies from U.S. federal R&D laboratories are also under way.

In cooperation with the Virginia Center for Innovative Technology (CIT), Virginia Tech has also assisted in the establishment of the Virginia Economic and Technology Development Program, a statewide technology-transfer network.

The University of Texas at Austin

The IC^2 Institute of the University of Texas at Austin is a major international research center for the study of Innovation, Creativity, and Capital—hence IC^2. The purposes of the IC^2 Institute are to:

1. Study, analyze, and report on the enterprise system through an integrated program of multidisciplinary research, conferences, symposia, and publications.

2. Provide unique approaches to the study of key business, economic, and technological issues. The institute deals with unstructured problems, develops multidisciplinary think teams, goes beyond functional boundaries in structuring projects, links theory with practice, provides opportunities to think anew within a university environment, and transfers research results to other institutions.

3. Bring together business, government, and academic leaders on critical issues through an active international program of conferences, workshops, and colloquia, and thus encourage a constructive exchange on policy initiatives and research activities.

4. Disseminate research findings through monographs, policy papers, technical working papers, journal articles, and book series with major publishing companies—directed to an audience that includes business people, academicians, students, government officials, and the general public.

IC^2 is a research organization that rigorously assesses the impact of the two key societal drives of technology and ideology; brings new analytical methods to bear on problems affecting the nation, individual states, local communities, academia, and business firms; and evaluates issues relating to the viability of emerging industries, the growth and survivability of business enterprises, and the role and purpose of private- and public-sector institutions.

RGK Foundation

The RGK Foundation in Austin, Texas, was established in 1966 to provide support for medical and education research. Major emphasis has been placed on medical research on connective tissue diseases, particularly scleroderma. The foundation also supports workshops and conferences at educational institutions through which the role of business in U.S. society is examined.

The RGK Foundation Building has a research library and provides research space for scholars in residence. The building's extensive conference facilities have been used to conduct national and international conferences. Conferences at the RGK Foundation are designed not only to enhance information exchange on particular topics, but also to maintain linkages among business, academia, community, and government.

About the Contributors

DAVID N. ALLEN is an Assistant Professor of Business Administration at the Pennsylvania State University, University Park, Pennsylvania.

UMBERTO BOZZO is the General Director of Tecnopolis CSATA Novus Ortus, Bari, Italy.

JOHN E. CANTLON is from the Office of the Vice-President for Research and Graduate Studies, Michigan State University, East Lansing, Michigan.

M. CAMPBELL CAWOOD is a General Partner of Venture First Associates, Winston-Salem, North Carolina.

THOMAS F. DOWD specializes in corporate and securities law and currently practices with Baker and Hostetler, Washington, D.C.

DEMETRIA GIANNISIS is Assistant Director of the University of Chicago's Technology Commercialization Center, Chicago, Illinois.

WILLIAM D. GREGORY is Professor of Physics at Clarkson University and Vice-President for Science and Technology of the DIOTEC Corporation, Potsdam, New York.

BRUCE R. HOPKINS specializes in tax-exempt organizations and currently practices law with Baker and Hostetler, Washington, D.C.

HERMAN E. KOENIG is Vice-President for Research Services and Industry Assistance at Michigan State University, East Lansing, Michigan.

HENRY C. KOWALSKI is Director of the Business and Industry Development Center at GMI Engineering and Management Institute (formerly General Motors Institute) in Flint, Michigan.

ILZE KRISST is Assistant Dean for Cooperative Research, Research Foundation, University of Connecticut, Storrs, Connecticut.

DOUGLAS H. MCQUEEN is a Professor at Chalmers University, Göteborg, Sweden.

NICHOLAS B. MAHER is Special Projects Assistant for the University of Chicago's Technology Commercialization Center, Chicago, Illinois.

THOMAS W. MAILEY is Project Director of the Ceramics Corridor in Upstate New York, a joint venture involving Alfred University, Corning Glass Works, and New York State.

D. BRUCE MERRIFIELD was, at the time of his writing the preface, the Assistant Secretary for Productivity, Technology, and Innovation, U.S. Department of Commerce, Washington, D.C. He is currently on the faculty of the Wharton School, University of Pennsylvania.

JAMES D. MORRISON is Associate Vice-President for Research and Professor of Chemistry at the University of New Hampshire, Durham, New Hampshire.

FREDERICK NORLING is an Associate Professor of Business and Economics at Muhlenberg College, Allentown, Pennsylvania.

ROMUALDO SABATELLI is the Assistant General Director of Tecnopolis CSATA Novus Ortus, Bari, Italy.

THOMAS P. SHEAHEN was most recently the Executive Director of the Energy Research Advisory Board of the Department of Energy. He is presently a private consultant.

STEPHEN SZYGENDA is a Professor of Electrical and Computer Engineering and the Director of the Center for Technology Development and Transfer at the University of Texas at Austin.

STANLEY T. THOMPSON is Administrative Coordinator for INventure, West Lafayette, Indiana.

DEBORAH WALKER is a law Partner in Peat Marwick's Washington National Tax Practice and specializes in employer benefits and executive compensation.

J. T. WALLMARK is a Professor at Chalmers University, Göteborg, Sweden.

WILLIAM E. WETZEL, JR., is Director of the Center for Venture Research and Professor of Business Administration in the Whittemore School of Business and Economics at the University of New Hampshire, Durham.

FRANK J. WILEM, JR. is President of the Gulf Coast Breeder in Pass Christian, Mississippi.

RAYMOND A. WILLIS is Director of the University of Chicago's Technology Commercialization Center, Chicago, Illinois.

DAVID I. WILSON specializes in licensing, intellectual property law, and trade law and he currently practices with Baker and Hostetler, Washington, D.C.

MEG WILSON is Coordinator of the Center for Technology Development and Transfer at the University of Texas at Austin.

About the Editors

ALISTAIR M. BRETT is Director of the Office of Technology Management and Transfer at the Virginia Polytechnic Institute and State University in Blacksburg, Virginia. He received a B.S. degree in physics from Kings College, University of London, and a Ph.D. in theoretical physics jointly from Drexel University and the University of St. Andrews, Scotland. Dr. Brett has taught at several colleges, and has served as a consultant to a number of small companies and also to Control Data Corporation's Business and Technology Centers Division.

Dr. Brett has assisted in developing Virginia Tech's technology-transfer program and has worked with the Virginia Center for Innovative Technology (CIT) to put in place a statewide technology-transfer and assistance program. He has worked to develop more efficient access to federal technology information, investigated federal laboratory spin-off companies for the Federal Laboratory Consortium for Technology Transfer, and assisted the Appalachian Regional Commission with new technology-transfer initiatives. Dr. Brett is the author and coauthor of several papers in theoretical physics and chemistry, and is a member of the Technology Transfer Society, the International Technology Institute, and the American Society for Training and Development.

DAVID V. GIBSON is an Assistant Professor in the Department of Management Science and Information Systems in the College and Graduate School of Business and a Research Fellow at the IC^2 Institute at the University of Texas at Austin. He received his B.A. degree from Temple

University, and M.A. degrees from the Pennsylvania State University and from Stanford University. In 1983 he earned a Ph.D. in sociology from Stanford after completing studies in the areas of organizational behavior and communication theory.

Dr. Gibson is co-director of the Annual Texas Conference on Organizations, which is sponsored by the Colleges of Business, Communication, and Education at the University of Texas at Austin, the RGK Foundation, and Price Waterhouse and includes organization scholars and graduate students from universities throughout Texas. Professor Gibson teaches undergraduate and graduate courses on communication behavior in organizations, international business, technology transfer, the management of information systems, corporate culture, and research methods. He belongs to the following professional associations: the Academy of Management, the American Sociological Association, and the International Communication Association.

Dr. Gibson's research and publications focus on the management of information systems, cross-cultural communication and management, and the management and diffusion of innovation. He is coeditor of *Creating the Technopolis: Linking Technology Commercialization and Economic Development* (Cambridge, Mass.: Ballinger Publishing, 1988), editor of *Technology Companies and Global Markets: Programs, Policies, and Strategies to Accelerate Innovation and Entrepreneurship* (Savage, Md.: Rowman & Littlefield, 1990), and coeditor of *Technology Transfer: A Communications Perspective* (San Mateo, Calif.: Sage, 1990).

RAYMOND W. SMILOR is the Executive Director and the Judson Neff Centennial Fellow at the IC² Institute, the University of Texas at Austin. He is also Associate Professor of Management in the UT Graduate School of Business and the Executive Vice-Chairman of the College of Innovation Management and Entrepreneurship of the Institute of Management Sciences.

Dr. Smilor has published extensively with refereed articles appearing in journals such as *IEEE Transactions on Engineering Management, Research Management, Journal of Business Venturing,* and *Journal of Technology Transfer.* His research areas include science and technology transfer, entrepreneurship, economic development, marketing strategies for high-technology products, and creative and innovative management techniques. His works have been translated into Japanese, French, and Russian.

He is a consultant to business and government. He has lectured internationally in China, Japan, Canada, England, France, and Australia.

He has also served as a research fellow for an NSF international exchange program on computers and management between the United States and the Soviet Union. He has been a leading participant in the planning and organization of many regional, national, and international conferences, symposia, and workshops.

Dr. Smilor speaks extensively to business, professional, and academic groups in the United States. He is also involved in several civic and professional organizations, and appears in *Who's Who in the South and Southwest*. He is editor or coeditor of nine books: *Corporate Creativity: Robust Companies and the Entrepreneurial Spirit* (New York: Praeger, 1984); *Improving U.S. Energy Security* (Cambridge, Mass.: Ballinger Publishing, 1985); *Managing Take-off in Fast Growth Companies* (New York: Praeger, 1986); *The Art and Science of Entrepreneurship* (Cambridge, Mass.: Ballinger Publishing, 1986); *Economic Development Alliances: Major New Relationships for Scientific Research and Technology Commercialization* (Austin: IC^2 Institute, 1987); *Technological Innovation and Economic Growth* (Austin: IC^2 Institute, 1987); *Creating the Technopolis: Linking Technology Commercialization and Economic Development* (Cambridge, Mass.: Ballinger Publishing, 1988); *Pacific Cooperation and Development* (New York: Praeger, 1988); and *Customer-driven Marketing: Lessons from Entrepreneurial Technology Companies* (Lexington, Mass.: Lexington Books, 1989).

He is coauthor of two books:

Financing and Managing Fast-growth Companies: The Venture Capital Process (Lexington, Mass.: Lexington Books, 1986); and

The New Business Incubator: Linking Talent, Technology, Capital and Know-how (Lexington, Mass.: Lexington Books, 1986).

Dr. Smilor has taught management and marketing courses at the UT College and Graduate School of Business. His course on "Technology Entrepreneurship" is rated as one of the best by students. He is also one of the most highly rated instructors in the Management Development Program in the UT Graduate School of Business. He earned his Ph.D. in U.S. history at the University of Texas at Austin.

Index